The Structure of Matter

from the blue sky to liquid crystals

André Guinier

Emeritus Professor in the University of Paris-Sud
Member of the Académie des Sciences

With a preface by Alfred Kastler, Nobel Laureate in Physics

Translated from the French by W. J. Duffin, Fellow of the University of Hull

Edward Arnold

We thank M. Winther for providing the link between the author of this book and its readers.

© Edward Arnold (Publishers) Ltd 1984

Authorized translation from the French
La structure de la matière du ciel bleu à la matière plastique
published by Hachette 1980

This translated edition first published in Great Britain in 1984 by Edward Arnold (Publishers) Ltd,
41 Bedford Square, London WC1B 3DQ

Edward Arnold, 300 North Charles Street, Baltimore, Maryland 21201, USA

Edward Arnold (Australia) Pty Ltd, 80 Waverley Road, Caulfield East, Victoria 3145, Australia

British Library Cataloguing in Publication Data

Guinier, André
    The structure of matter.
    1. Matter—Construction
    I. Title   II. La structure de la matière du ciel bleu à la matière plastique. *English* 539'.1   QC173

ISBN 0-7131-3489-5

Text set in Photon Times and Univers by SB Datagraphics Ltd
Printed and bound by Spottiswoode Ballantyne Ltd, Colchester and London

Preface

The structure of the matter which forms the world around us and even our own bodies has fascinated the human intellect from the earliest times.

For some, the structure had to be continuous: after all, does not a typical substance like water have the same properties at every point, and does it not seem to remain homogeneous however finely it is divided, whether in practice or in our imagination? For others, it was quite the opposite: is matter not discontinuous on so fine a scale that it is imperceptible to us even with the help of a microscope, and is it not formed of distinct and countable particles which the ancient Greeks themselves called atoms and molecules? Moreover, if the second group, the visionaries, are right, on what submicroscopic scale does the discontinuity appear?

This controversy over the two concepts of matter has endured for many centuries and has continued to fascinate scientists and philosophers. It was particularly vigorous during the second half of the nineteenth century when analysis of the chemical constitution of substances led to the establishment of the quantitative laws of chemistry and showed that the atomic theory accounted admirably for all the facts. However, at that point chemists came up against the scepticism of positivist physicists who were disciples of Auguste Comte, and who were trying to eliminate all models in science that were supposed to underlie observable reality. They insisted that the sole aim of science was to establish relations between sensory perceptions and that everything not 'observable' should be excluded from scientific thought. Duhem in France, Ostwald in Germany and Ernst Mach in Austria were the fiercest defenders of this point of view: they made life hard for the Viennese physicist Ludwig Boltzmann, who was championing the ideas of the atomists.

The situation changed completely with the advent of the twentieth century: not only do we now know how to count the number of atoms in a given piece of matter and to measure their dimensions, but the physicist of today is undertaking the exploration of the internal structure of atoms so that they no longer merit their name of the 'indivisible ones'.

The first breakthrough was made as long ago as 1865 by the Austrian physicist Joseph Loschmidt, a contemporary and friend of Boltzmann, who first determined the order of magnitude of atomic sizes and showed that molecular dimensions were somewhere around one ten-millionth of a millimetre.

The Italian physical chemist Avogadro had already shown at the beginning of the nineteenth century that, if molecules existed at all, then the same volume of any gas at the same temperature and pressure must contain the same number of molecules whatever the nature of the gas. Quite independently, André-Marie Ampère had arrived at the same conclusion, and chemists had become accustomed to talking of Avogadro's number N: the number of molecules contained in a 'mole' of the substance, whether gas, liquid or solid.

How did Loschmidt manage to determine the order of magnitude of this number N? It was through the kinetic theory of gases, introduced by Daniel Bernouilli from the eighteenth century and developed towards the middle of the nineteenth century by Clausius and Maxwell. This theory considered a gas as a swarm of molecules moving randomly in space and colliding continuously with each other. Accordingly, the internal energy of a gas was the sum of the kinetic energies of its molecules. The pressure exerted by a gas on a wall was interpreted as the result of the forces produced by the molecules in rebounding from it. Finally, the viscosity of the gas—determined by measuring its rate of flow through a capillary tube—had its origin in the forces exerted by the molecules on each other during collisions. Clausius and Maxwell had succeeded in expressing the coefficient of viscosity of a gas η as a function of Avogadro's number N and of the diameter d of the molecules (assuming them to be hard spheres, although this was clearly only a rough approximation). The formula can be written

$$\eta = 0.53 \frac{\sqrt{(MRT)}}{\pi N d^2}$$

where M is the molar mass and R the universal gas constant (see p. 50 of this book for another form of this expression). Measurement of the coefficient η yielded a relation between the two unknowns N and d, but to determine them separately needed a second relation. To obtain that, Loschmidt assumed that he could compare the state of condensed matter (solid or liquid) to that of a heap of spheres of diameter d in contact with each other. The molar volume V of a pure substance is then given by a formula of the type

$$V = \alpha N d^2$$

where α is a numerical coefficient close to unity depending on the symmetry and closeness of the packing. Since it was simply a question of obtaining some idea of the order of magnitude of molecular dimensions, Loschmidt contented himself with the approximation $V = N d^3$ which he applied to the liquefied gases O_2, N_2 and CO_2 (a few years previously, Cailletet had succeeded in liquefying the first two of these gases). The result of this calculation enabled Loschmidt to assert that the number N must be between 10^{22} and 10^{24} and that the diameter of simple molecules had to be of the order of 0.1 to 1 nanometre. We now know N with great precision because it has been determined by some dozen different and independent methods. It is given by

$$N = 6.022 \times 10^{23} \text{ mole}^{-1}$$

We must emphasize the enormous size of N. The number of molecules in each cubic centimetre of air around us is of the order of 3×10^{19}. If these molecules could be counted at the rate of one per second the enumeration would take 3×10^{19} seconds, a time which should be compared with the estimated age of the universe: 10 000 million years or a mere 10^{17} seconds!

As for the molecules, we now know not only their sizes, but also their shapes, their internal structures and the forces exerted between them: all subjects to be discussed in this book.

One of the most elegant methods for determining N was through the study of Brownian motion by Jean Perrin at the beginning of this century. This random perpetual agitation of fine particles suspended in a liquid is normally seen only under an ultra-microscope, but it forms one of the most striking demonstrations at the *Palais de la Découverte* in Paris where the microscopic image of the motion is thrown on to a television screen.

Another method for determining N, perhaps even more extraordinary, is through the measurement of the intensity of blue light from the sky, light which is the result of the scattering by air molecules of sunbeams passing through the terrestrial atmosphere. In a treatise on colour, Goethe had carefully noted the behaviour of a beam of white light passing through a cloudy medium. The beam that emerged appeared reddish, while observation of the beam from the side against a dark background as it passed through the body of the medium showed a dark blue colour. Goethe had found a correspondence between that phenomenon and the colours of the setting sun and day-time sky. However, it was not until 1900 that the English physicist Lord Rayleigh made it clear that it was not only suspended particles that could scatter light but that molecules themselves could also do so, and that the scattered intensity enabled the number of molecules in the gas to be counted.

So contemplation of the blue sky—the inspiration of so many poets—has enabled the physicist to say that the colour gives us daily proof of the discontinuous structure of the scattering matter: blue light is more strongly scattered than red because its wavelength is closer to the dimensions of a molecule. If atmospheric air were a continuous medium—in other words, homogeneous at any submicroscopic level however fine—the sky would be black and we should be able to see the stars in full daylight as cosmonauts do outside our atmosphere.

The atomic and molecular theory celebrated its greatest triumphs in the field of solids, particularly with crystals, whose regularity of form had fascinated people from the beginning of time. René-Just Haüy in the eighteenth century had already sensed that this macroscopic regularity must be due to a regular submicroscopic structure having three-dimensional order arising from a packing together of what we call today the 'elementary unit cell'. A striking proof of the accuracy of this idea was provided in 1912 when Max von Laue, then a professor at the University of Munich, produced what we could call the 'Columbus's egg' of the twentieth century. He suggested the diffraction of X-rays by a crystalline lattice, an idea both simple and inspired that has yielded such a rich harvest today. Nobody is better qualified than the author of this book to present the results of that epic story, in which he has personally taken part and to which he has made such important contributions.

Between the perfect order exhibited by a silicon crystal and the complete disorder of a gas at low pressure, there is an infinite variety of structures shown by matter in many forms. It is the great merit of this book that it reveals the essential features of these structures without the use of any mathematical formalism, and that it also shows us the importance of knowledge at the molecular level for the many techniques developed by the human race over the ages. Thus, it throws light on the 'why?' of techniques that go back to earlier times (the metallurgy of bronze and steel, the art of pottery), but even more it shows how a knowledge of structures enables us to develop new techniques such as the manufacture of plastics and synthetic fibres, some of which have strengths greater than steel together with a remarkable flexibility at the same time. Almost every day, materials are being developed with novel properties which contribute to advances in the fields of electronics, computers, biology and medicine. Lives are already being prolonged through the use of artificial hearts and kidneys. Human ingenuity, that has made us masters of such a profound knowledge of matter, will undoubtedly still provide us in future with many surprises, as long as we learn how to use scientific advances for the good of us all and how to prevent them from being used for harmful purposes: let there be no doubt that this is the fervent wish of those taking part in the great adventure of science.

A. Kastler

Foreword

I should like to thank M. A. Kastler very warmly for being kind enough to write the preface to this work. He has recalled some of the decisive events in the unravelling of the structure of matter, but has adopted a point of view that will not be found in the following pages. This is because, when undertaking to deal with so vast a subject in a book of strictly limited size, a number of preliminary choices must be made.

The central theme of the book is a description of 'models' that all physicists and chemists now accept and use constantly in their fundamental as well as their applied research. I have, however, deliberately chosen not to justify these models: in other words, not to deal with the experiments and theories which have contributed to their construction.

I have chosen, too, not to limit myself to simple cases, such as the perfect gas, crystals and so on that are favoured in standard textbooks, but to talk about complicated systems as well, systems that we see in our normal surroundings or that we use everyday, such as plastics, rubber, concrete, even mayonnaise. ... These excursions have unhappily prevented me from developing certain chapters in the original plan which dealt with the relation between the structures of solids and the properties that can be deduced from them, such as thermal, electric, magnetic or mechanical properties. This would provide the subject matter of a second volume in the same spirit as this one.

I have also chosen to use no mathematical formalism. That does not imply any lack of precision, but rather that the description of models as well as their possible uses is deliberately incomplete. However, there are so many books at various levels that make normal use of mathematics that it was tempting to show what could be done without it. This book can be used in parallel with more standard works either as an aid in using them or by suggesting ideas to those who take too much pleasure in merely manipulating formulae.

A level of knowledge equivalent to that of the present Terminales [the final year in French lycées where students are prepared for higher education] is sufficient to embark on this work. It is quite pointless, however, to attempt to define the level of a book too closely, for a passage that interests one reader will be judged banal and superfluous by another and difficult to understand by a third. In any case, I dare to hope for quite a varied readership. Authors should consider themselves fully recompensed for their efforts if *all* readers could draw from their work even a *few* ideas that were useful to them.

There is, however, one category of reader that I have constantly kept in mind while writing: the physical science teachers from the Sixième to the Terminale [Sixième = first year of secondary education normally at eleven years old]. Many ideas have recently been introduced in new programmes and some teachers have not had an initial training which might suit them for their current tasks. For teachers to achieve their proper goal, that is, to open up the minds of pupils to the external world and not to rely on statements of a few formulae, it is essential that they have a level of knowledge clearly above the one they have to teach. It is to help teachers in such efforts that this book has in large measure been conceived, for the study of the structure of matter holds an important place in the programmes of all classes of student. Obviously, however, I hope that my work will be useful to other teachers and to non-teachers.

A book of this sort cannot be weighed down by references and indications of original sources. For that reason, I should like here to do justice to all the authors whose works I have used but not cited and to express my sincere thanks to the numerous colleagues who have given me information or permitted me to use their papers. The list is so long that it could not be given without omissions, except for the photographs, which are reproduced with the names of their authors.

I am delighted to thank especially MM. R. Omnes and H. Gié, general editors of the series* that my book has the honour to inaugurate, for their encouragement and comments. I am also very indebted to M. M. Winther for having judiciously criticized the manuscript from the point of view of a teacher. Finally, I am happy to say how much I have valued the help of my friend Ch. Mazières who took an active part in the preparation of this work.

A. Guinier

* This book is the translation of a volume in the series 'Liaisons Scientifiques' published by Hachette under the general editorship of Roland Omnes and Hubert Gié.

Contents

Mendeleev's periodic table

Pointer labels:
- principal quantum number of the outer electron shell
- maximum possible number of electrons in that shell
- atomic number, Z
- chemical symbol of the element
- atomic mass

n	shell	max. electrons		1	2	3	4	5	6	7	8	9	10	11	12	13	14	15	16	17	18
1	(K)	1	2	1 **H** 1																	2 **He** 4
2	(L)	4	8	3 **Li** 6.9	4 **Be** 9											5 **B** 10.8	6 **C** 12	7 **N** 14	8 **O** 16	9 **F** 19	10 **Ne** 20.2
3		18		11 **Na** 23	12 **Mg** 24.3											13 **Al** 27	14 **Si** 28	15 **P** 31	16 **S** 32	17 **Cl** 35.5	18 **Ar** 40
4		32		19 **K** 39.1	20 **Ca** 40	21 **Sc** 45.1	22 **Ti** 47.9	23 **V** 51	24 **Cr** 52	25 **Mn** 54.9	26 **Fe** 55.8	27 **Co** 58.9	28 **Ni** 58.7	29 **Cu** 63.8	30 **Zn** 65.4	31 **Ga** 69.7	32 **Ge** 72.6	33 **As** 74.9	34 **Se** 79	35 **Br** 80	36 **Kr** 83.7
5		50		37 **Rb** 85	38 **Sr** 87.6	39 **Y** 88.9	40 **Zr** 91.2	41 **Nb** 92.9	42 **Mo** 96	43 **Tc** 99	44 **Ru** 102	45 **Rh** 103	46 **Pd** 107	47 **Ag** 108	48 **Cd** 112	49 **In** 115	50 **Sn** 119	51 **Sb** 122	52 **Te** 128	53 **I** 127	54 **Xe** 131
6		72		55 **Cs** 133	56 **Ba** 137	57 **La** 139	72 **Hf** 179	73 **Ta** 181	74 **W** 184	75 **Re** 186	76 **Os** 190	77 **Ir** 193	78 **Pt** 195	79 **Au** 197	80 **Hg** 201	81 **Tl** 204	82 **Pb** 207	83 **Bi** 209	84 **Po** 210	85 **At**	86 **Rn** 222
7		98		87 **Fr**	88 **Ra** 226	89 **Ac** 227															

[14 rare earths or lanthanides]

90 **Th** 232	91 **Pa** 231	92 **U** 238	[transuranic elements →]

Elements commonly found on the earth

Elements forming more than 0.1% of the earth's crust: O, oxygen; Si, silicon; Al, aluminium; Fe, iron; Ca, calcium; Na, sodium; Mg, magnesium; K, potassium; Ti, titanium; H, hydrogen; P, phosphorus; Mn, manganese.

Other elements forming more than 1% of the atmosphere: N, nitrogen; Ar, argon.

Element forming more than 1% of the oceans: Cl, chlorine.

1 The basic elements of structures: atoms, molecules, ions

Introduction

All matter is composed of atoms, and there is a different atom corresponding to each chemical element: these are facts which are now universally accepted and known to everybody. (Mendeleev's periodic table on the opposite page gives a list of all the chemical elements and it can be seen that those commonly found on the earth are not very numerous.)

When we talk about establishing a model of the structure of a substance, we mean 'describing the arrangement of atoms in it' or, in other words, identifying every atom and specifying its respective position. The easiest way of looking at that would be to regard each atom as a perfectly definite and unchangeable constituent of the substance and as possessing its own specific properties. However, the real situation is not as simple as that and some minor modifications are needed if we are to account for the properties of matter. There are three ideas that must be introduced:

1. The first is the structure of an atom itself with its nucleus and Z electrons, a picture that should be familiar to the reader since it occurs at the beginning of many science courses. Although quantum theory has added refinements to this model, we shall not need them. The one indispensable idea is the distinction between the heart of the atom, its nucleus, and the electrons around it.

Each of the Z electrons of an atom is in a 'state', characterized among other things by the energy it would take to detach it from the atom without giving it an appreciable speed—this is called the **binding energy** of the electron. The values of these energies allow the Z electrons to be grouped into 'shells', each with its own distinct energy (or small range of energies) well separated from others. *The number of electrons that can possibly be accommodated in each shell is limited.* Thus, the deepest shell of every atom (called the 1s or K shell) can only accommodate two electrons: these are bound most strongly of all to the nucleus and are situated closest to it. Their binding energy is greater than that of the other electrons, varying from one chemical element to another roughly as Z^2, and it ranges from 13.6 eV[1] for hydrogen to 115 600 eV for uranium ($Z = 92$). The most loosely-bound electrons, however, are those in

---

[1] The electron-volt (eV), the most commonly-used unit of energy in atomic physics, is the energy gained by an electron when falling through a potential difference of 1 volt. One eV per atom corresponds to a macroscopic (large-scale) unit of 96.5 kJ per mole.

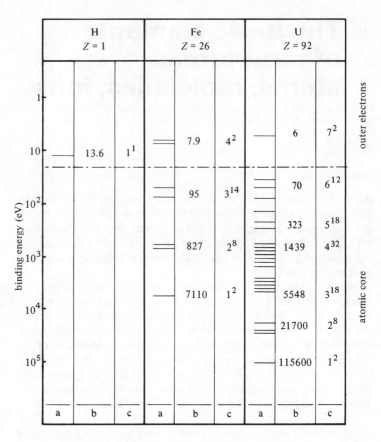

Fig. 1.1 Binding energies of the electrons in various shells for three atoms, with small, medium and large atomic numbers respectively: hydrogen, iron and uranium. For each atom we have in column a, representation of the levels of binding energies of the various electrons in the shells; in column b, maximum value of binding energy in the shell; in column c, principal quantum number of the shell, with an exponent giving the number of electrons it contains.

the outermost shells, and for all atoms these have binding energies of the order of only a few eV (see Fig. 1.1).

In a physical change or a chemical reaction, only the states of a few outer electrons of the atoms are changed, because the amounts of energy involved are so much smaller than the binding energies of the inner electrons. So cooks in their kitchens, chemists in their laboratories, even metallurgists in their blast- and arc-furnaces, can only scratch the outer layers of atoms. To a very good first approximation, the **core of the atom**—that is, *the nucleus plus all but a few of the electrons in the outermost shells*—remains unchanged. It follows that the nucleus[1] itself is certainly not affected, because to alter that would require energies around one million eV, far outside the normal physico-chemical values.

2. The second important idea is that all matter is made up from a collection of 'entities', just as a building is composed of bricks. In matter, however, these entities can be of different types, so it is convenient to call them in general 'particles' or 'basic particles of the model'. In a few rare cases, the basic particles can be isolated single atoms. Rather more often they are still single atoms but

[1] Remember that the word 'atom' in common usage (in economics, ecology, etc.) is in fact often used to describe the atomic nucleus.

with their outer electron shells modified. However, most frequently of all they are definite *groups of atoms*, or **molecules**.

3. The third point is that the basic particle can be represented by an object of definite shape and size, and that the model of the material is the whole collection of these objects. A single atom, for instance, can be represented by a sphere of given diameter, and a collection of atoms by a set of spheres in contact with each other. Although such a model is not necessarily obvious, its validity has been well tested by experiment and it greatly simplifies our task of understanding the structure of matter.

All that we have said so far in this introduction explains why the main aims of the first chapter are, firstly, to describe the *nature* of the basic particles and, secondly, to look at their *sizes*.

The structural elements of models: monatomic and polyatomic molecules

First of all, we consider the only really simple case: that of the inert gases (He, Ne, Ar, etc.). The atoms of these substances all have a *completed outer electron shell*, that is, one which cannot accept another electron. Such a complete shell is particularly stable: thus, to detach an electron from it requires an energy (the ionization energy) greater than that needed to ionize other atoms—between 10 and 25 eV instead of just a few eV. Although a very few compounds of these gases are known, in practice they generally behave as inert substances and their atoms do not interact with neighbouring ones, whether identical or different. Thus, the materials called 'inert gases' do consist of a collection of atoms each identical to a single isolated atom.

The six inert gases (helium, neon, argon, krypton, xenon and radon) are the only substances in which the basic particle is a single unchanged atom. Apart from those, every type of atom interacts with any other placed in contact with it, the interaction being due to a change in the state of some of the electrons in the outer shells.

Descriptions of the various types of binding and of the way they arise belong to the enormous field of chemistry and we shall certainly not want to go very far into such matters. We shall simply emphasize the existence of **molecules** and look a little more closely at the changes which take place in atoms when they interact with others: we concentrate on these two aspects because our aim is mainly to catalogue the basic particles used in models of the structure of matter.

A molecule is a group of atoms bound to each other in a well-defined and stable arrangement. The stability is great enough for its structure to remain unchanged during physical transformations such as melting, boiling, etc. A complete description of a molecule needs, firstly, a diagram of the structure giving the positions of the cores of the constituent atoms and, secondly, a specification of the states of the electrons which contribute to the bonds between atoms.

H—H   HC ≡ CH   CH₃—CH₂—CH₂—CH₃

H_2 C_2H_2 C_4H_{10}

hydrogen   acetylene   butane

$M = 2$ $M = 24$ $M = 58$

$$CH_2{-}OH$$
$$|$$
$$CH{-}OH$$
$$|$$
$$CH_2{-}OH$$

$C_3H_8O_3$

glycerol

$M = 92$

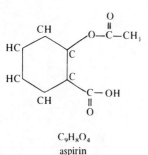

$C_9H_8O_4$
aspirin
$M = 180$

$C_{16}H_{24}O_3N_2S$
penicillin G
$M = 324$

Fig. 1.2 A few examples of small- and medium-sized molecules. *M* is the molecular mass using the H atom as the unit.

Both in pure chemical elements and in compounds, the basic particles are very frequently, but not always, molecules. Thus, the materials called 'hydrogen' and 'oxygen' are made up of molecules H_2 and O_2 respectively, the material 'water' of molecules H_2O, and so on. Even the inert gases could be included here by considering their molecules to be *monatomic*, all others being *polyatomic*.

As well as the very simple molecules just quoted as examples, there are others which become progressively more complex (see Fig.1.2). From a few atoms per molecule, we pass continuously through to a few dozen, a few hundred, and then to *macromolecules*, with more than 10 000 constituent atoms, such as those encountered in molecular biology. As an example, the molecule of *insulin* contains about 1500 atoms (of H, C, N, O, S) and its molecular mass *M* is around 12 000 (in terms of H = 1 or O = 16). Its structure is now known exactly and Fig. 1.3 gives some idea of its complexity.

From the simplest diatomic molecules to the giant ones in their infinite variety, there is one common factor: each forms a well-defined assemblage which can be neither divided, nor deformed, nor increased in size by growth, without using an appreciable amount of energy. The exact arrangement of atoms in molecules and their stability can be explained by the nature of the electronic bonds and we shall return to this in Chapter 4.

Some molecules are more easily disrupted than others, but all of them have enough stability to be the *permanent* 'particles' which make up a substance. Thus, in a gas, although the molecules are not subjected to forces for most of the time, each one does suffer around 10^{10} collisions per second with the walls of the containing

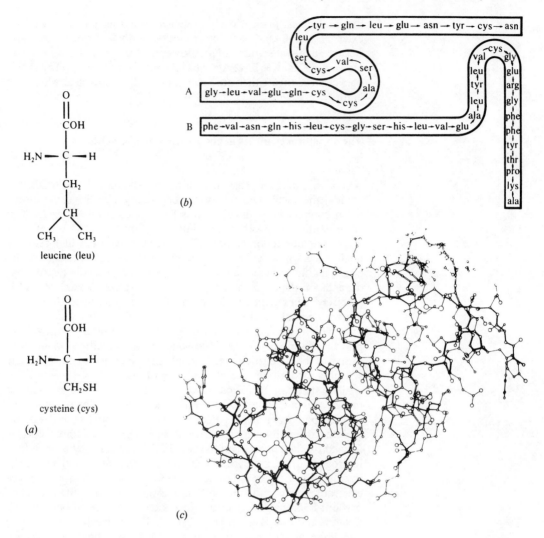

(a)

O
‖
COH
|
H₂N ── C ── H
|
CH₂
|
CH
╱ ╲
CH₃   CH₃

leucine (leu)

O
‖
COH
|
H₂N ── C ── H
|
CH₂SH

cysteine (cys)

(b)

A: gly→leu→val→glu→gln→cys

B: phe→val→asn→gln→his→leu→cys→gly→ser→his→leu→val→glu

tyr → gln → leu → glu → asn → tyr → cys → asn

leu, ser, cys, val, ser, ala, cys

cys, gly, leu, glu, tyr, arg, leu, gly, ala, phe, phe, tyr, thr, pro, lys, ala

(c)

Fig. 1.3 The structure of a giant molecule: insulin from cattle. The constituent elements are amino-acids. (a) Two examples of amino-acids: the formulae of leucine and cysteine. (b) The two chains of insulin, A and B, containing 21 and 30 amino-acids respectively. (c) The structure of the molecule as it exists in an insulin crystal: the chains are curled up in space and cross-linked by bridging atoms.

vessel or with other molecules, but these collisions do not destroy them or even deform them. However, they can be changed by receiving sufficient energy from outside. For example, air molecules at a great height are ionized by the action of ultraviolet solar radiation or, to put it another way, the collision of a photon with a molecule can detach an electron from it. (This is the origin of the conducting layer in the atmosphere which plays an essential role in the propagation of radio waves round the earth: the *ionosphere*.) Another example: the hydrogen molecule, H_2, begins to dissociate into two H atoms above 2500 °C and the dissociation is practically complete at 6000 °C; the H atoms are unstable at lower temperatures and immediately recombine into H_2 molecules. (This has an application in atomic hydrogen welding.)

When matter is in a rarified state (i.e. a gas), the basic particles are, as a general rule, molecules, but this is only *sometimes* true of matter in a condensed state, particularly in solids. If it *is* true, then the molecules pack closely together like rigid bodies with an external shape and size determined by their atomic structure. Beside such *molecular solids*, however, there are those in which molecules, as we have defined them, *do not exist*. The basic particles are then either **atoms** or **ions**, and we now deal with these in turn.

The structural elements in condensed states of matter

Atoms

Although atoms can certainly form the basic particles in solids and liquids, they will then no longer be exactly like the same atoms when completely isolated. This is because each one finds itself in close contact with others and interacts with them. As we already know, the interaction cannot affect the atomic core, which remains unchanged whatever the other atoms may be, but the state of the electrons in the outer shells can be greatly altered, to a degree that depends on the environment of the atom. As we also know, that is precisely what happens to atoms forming molecules, where the essential feature is the formation of a well-defined finite system to which other atoms cannot be added or subtracted without profoundly modifying it. However, when the atoms *themselves* form the basic particles of a solid, they interact to form chains stretching from one side of the block of material to the other. We shall be studying the formation of bonds between atoms in a solid in more detail in Chapter 4, but let us fix ideas here by an example.

A carbon atom has six electrons, two in the first shell (1s or K) and the other four in the second shell (L). The atomic core is formed by the nucleus and the two 1s electrons. Together with four hydrogen atoms, the carbon can form a *molecule* of methane, CH_4, in which the C atom is bound to the four H atoms by four pairs of electrons, one each from the H atoms and four from the L shell of carbon. Each pair of electrons forms a *covalent bond* between the C atom and one H atom (Fig. 1.4(a)). The molecule CH_4 is stable and finite in the sense that another H atom cannot be stuck on (CH_5 doesn't exist), while to detach a H atom needs an energy input of 4 eV and leaves behind a CH_3 'radical' which is so unstable that its mean lifetime, before recombining with a H atom, is of the order of 10^{-6} second.

By contrast, in diamond, an atom C_0 as in Fig. 1.4(d) is also linked by covalent bonds to four neighbours, but this time they are *carbon* atoms. Each bond is formed by a pair of electrons from the L shells of two neighbouring C atoms, C_0 and C_1. The atoms C_1 are at the apexes of a regular tetrahedron, just like the H atoms in methane. Unlike the H atoms, however, the C_1 atoms are also linked to four neighbours (for example, the three C_2s and C_0) and this continues throughout the solid. The C atoms in diamond form chains (Fig. 1.4(c)) which cross each other and can continue indefinitely in all directions.

Diamond is a typical example of a solid (a crystal) in which the basic particles are *atoms*—**there are no molecules**. Nevertheless,

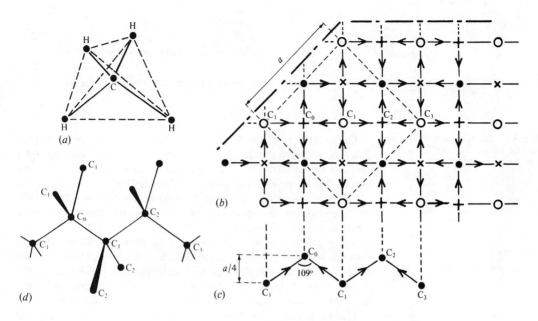

those basic particles are not identical with free carbon atoms since the electrons in their L shells are in a different state that is produced by their neighbours. In methane, all the bonds of the carbon and hydrogen atoms are satisfied, or *saturated*, and this results in a finite molecule. In diamond, however, the structure can extend indefinitely and could perhaps be considered as a molecule of infinite extent. In fact, of course, the real solid is necessarily limited in size, so that the carbon atoms at the surface of a block would have bonds remaining free. That would not be a stable state of affairs and it is very likely that the surface atoms would combine with impurities from the surrounding atmosphere. Even if a diamond is placed in the highest vacuum available, there are always enough residual atoms around to cover the surface. Whatever form it takes, the surface of a solid clearly has a slightly perturbed atomic structure, although this has no effect on the internal structure because surface atoms are such a small proportion of the total (for instance, for a very small one carat diamond, 0.2 g, in the form of a cube, only ten atoms in every million are in the surface). The study of the atomic structure of solid surfaces, known as *surface physics*, is a recent development, but is currently a very active field of research because of its theoretical interest and technological applications.

Fig. 1.4 (a) The structure of the methane molecule, CH_4. (b) The structure of diamond. This represents a three-dimensional arrangement projected on to the page. The repeating unit, or unit cell, is a cube of side a. Atoms ○ are at zero height; atoms ●, at height $a/2$; atoms +, at height $a/4$; atoms ×, at height $3a/4$. Two neighbours are linked by a bond shown projected as a vertical or horizontal line, but which in fact rises by $a/4$ in the direction of the arrow. (c) The arrangement in an infinite chain of C atoms. Diamond can be considered as constructed from such chains crossing each other. (d) Each atom is surrounded by a tetrahedron of four other C atoms. For the two neighbours C_0 and C_1, the bonds C_0C_1 and C_1C_2 are parallel but in *opposite directions*.

Ions

An atom whose atomic number exceeds that of any inert gas by 1, 2, 3 or 4, and which loses its outer electrons, is left with the same electronic configuration as the inert gas. The removal of the outer electrons requires only a small amount of energy and so can take place quite readily, but it leaves the atom no longer electrically neutral: it has become a *positively-charged ion* with 1, 2, 3 or 4 elementary charges.

$$\boxed{Na} + 5.14\,eV \rightarrow Na^+ + e^- \qquad\qquad\qquad \text{ionization: absorption of energy}$$

$$\boxed{Cl} + e^- \rightarrow Cl^- + 3.71\,eV \qquad\qquad\qquad \text{affinity of Cl for an electron: liberation of energy}$$

$$Cl^- + Na^+ \rightarrow \boxed{Cl^-\,Na^+} + 7.9\,eV \qquad\qquad\qquad \text{electrostatic attraction: liberation of energy}$$

Result

$$\boxed{Na} + \boxed{Cl} \rightarrow \boxed{Cl^-\,Na^+} + (7.9 + 3.71 - 5.14 = 6.5\,eV) \quad \text{formation of the molecule: liberation of energy}$$

Fig. 1.5 The transition from two isolated atoms, Na and Cl, to the ionic molecule (Na^+Cl^-).

Conversely, consider an atom with 1, 2, etc., electrons *fewer* than those of an inert gas so that there are vacant places in the outer shell. A free electron from outside the atom would fall to a lower energy if it occupied such a vacant place, and in this way the atom tends to complete its outer shell spontaneously until it achieves the configuration of the inert gas. In this case, the atom becomes a *negatively-charged ion*.

Some atoms can also be transformed into positive or negative ions by losing or gaining electrons even without their outer shells becoming exactly like those of inert gases. This is possible when the new configuration happens to possess exceptionally high stability. Examples of this type are Ag^+, Cu^{2+} (see Mendeleev's periodic table).

The formation of a positive and a negative ion can take place simultaneously. For instance, if *one* atom of sodium is in contact with *one* atom of chlorine, a reaction takes place which can be broken down into three parts, as in Fig. 1.5:

(1)   The Na atom becomes a Na^+ ion.
(2)   The liberated electron transfers to the Cl atom which turns into a Cl^- ion.
(3)   The Na^+ and Cl^- ions are bound together by the usual force of electrostatic attraction between charged bodies of opposite sign (Coulomb's law). A NaCl molecule is formed with an ionic bond between the Na^+ and Cl^-: this process is possible because the three separate stages taken together lead to an overall liberation of energy.

A true NaCl molecule does exist in the gaseous state, but solid—and liquid—sodium chloride consist of a collection of positive and negative ions in such proportions that the total is electrically neutral as in all ionic solids and liquids (for instance, one single charge +ve ion to each single charge −ve ion, one double charge +ve ion to every two single charge −ve ions, etc.). In addition, the arrangement of the two sorts of ion, positive and negative, is such that neutrality is assured even over small regions containing only a few ions. There is an alternation of charges of opposite sign which leads to great cohesion as a result of electrostatic attraction. In such arrangements, there are no isolated positive–negative pairs and so *there are no molecules*. The basic particle in this model is the *ion*, that is, an atom whose outer shell has been very simply modified by the loss or gain of a small number of electrons.

Just as in solids with covalent bonds, the effect of the external surface of the material is very small and the total number of ions is not affected by it.

We have talked of an ion as being produced from one atom, but there do exist more complex ones. These are composed of several atoms bound to each other as in a molecule but with the total charge of the electrons not exactly compensated by that of the nuclei. In short, they are charged molecules whose electrons have a particularly stable configuration, thus allowing them to behave as would a permanent particle. Figure 1.6 gives three examples of such complex ions.

The sizes of ions, atoms and molecules

Having now catalogued the basic 'particles' used in models of structures of materials, we now go on to describe some of their properties which we shall need for further discussion of the models.

Their *mass* presents no special problems since it is essentially concentrated in the nuclei. For most atoms, the electrons contribute about 1/4000 of the total mass, so that small changes in their numbers or modification of their states produce no significant variations. The mass of the 'particle' is thus simply that of the nuclei in it.

Geometrical factors, such as *size* and *shape*, are rather more complicated. The quantum mechanical picture of an *isolated atom* consists of a cloud of negative electric charge of total amount $-Ze$ around a point nucleus of charge $+Ze$. The density of the negative charge at any point can be calculated from the fundamental equations of quantum mechanics and its value is found to decrease with distance, becoming zero at infinity. An isolated atom, therefore, has no precise limits. The hydrogen atom is one example that can be rigorously calculated and the results are shown in Fig. 1.7 by giving the radii of spheres containing respectively 50%, 90%, 99% and 99.9% of the electronic charge.

The situation is different in solids and liquids where the atoms are densely packed. The simplest case is that of **ions**. We start from an experimental fact of great importance in the establishment of atomic structures: in ionic crystals, formed by alternating ions of opposite signs, the distance between neighbours can be measured with great precision, as we shall see in Chapter 4. By collecting together such distances for a large number of different crystals, but involving only a limited number of different ions, we arrive at the following fundamental rule: *to a good approximation, each type of ion can be represented by a* **hard sphere** *of given radius*. The term 'hard sphere' means that it does not deform when coming into contact with its neighbours.

It is then found that the distance between the centres of two neighbouring ions in *any* crystal can be calculated by adding the radii of the two ions as if they were two rigid spheres in contact. In practice, the rule is not quite exact, but the difference between measured and theoretical values is always very small, usually less than 2%: it is a very reliable guide which is used again and again to select *possible* structures for crystals out of the many that can be put forward.

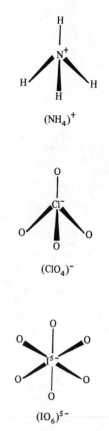

Fig. 1.6 Configurations of several complex ions. Ammonium ion, $(NH_4)^+$: the N atom is surrounded by eight electrons, five coming from the N and three from the four Hs, leaving one positive charge on one H atom unbalanced. Perchlorate ion, $(ClO_4)^-$: the Cl atom is surrounded by eight electrons, seven coming from the Cl itself and one extra electron which gives the whole ion its negative charge. Each O atom is also surrounded by eight electrons, six from the Os themselves and two being common to an O and a Cl (covalent bond). Periodate ion, $(IO_6)^{5-}$: the total number of valence electrons is 48, 36 coming from six Os, seven coming from the I, with five extra electrons giving the ion its total charge. The six Os are each surrounded by eight electrons.

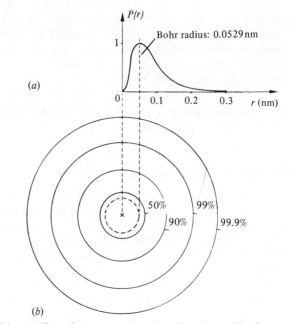

(a)

(b)

Fig. 1.7 Distribution of the electron in the hydrogen atom in its ground state according to quantum mechanics. (a) The probability of finding the electron at a radial distance r from the nucleus. $P(r)\,dr$ is the probability of finding it between r and $(r + dr)$. $P(r)$ is a maximum at the Bohr radius $r_0 = 0.0529$ nm. (b) Cross-sections through spheres containing various fractions of the electronic charge on the average. The fraction is 50% for $r = 0.0707$ nm, 90% for $r = 0.1400$ nm, 99% for $r = 0.225$ nm and 99.9% for $r = 0.297$ nm.

Table 1.1 lists the accepted values for the radii of a number of ions, and Table 1.2 gives several examples showing the comparison between *measured* interatomic distances in crystals and *calculated* values using the radii of Table 1.1.

The sizes of ionic radii are not merely raw experimental data: they are also related in a simple way to the electronic configuration of the ions. Consider first the various groups of ions which have the same configuration as a given inert gas (see Mendeleev's periodic table):

(He): Li^+, Be^{2+}
(Ne): O^{2-}, F^-, Na^+, Mg^{2+}, Al^{3+}, Si^{4+}
(Ar): S^{2-}, Cl^-, K^+, Ca^{2+}, Sc^{3+}, Ti^{4+}
(Kr): Se^{2-}, Br^-, Rb^+, Sr^{2+}, Y^{3+}, Zr^{4+}
(Xe): Te^{2-}, I^-, Cs^+, Ba^{2+}, La^{3+}, Ce^{4+}

In each of these groups, the number of electrons is the same while the nuclear charge increases from the first chemical element to the last. Each electron is therefore attracted by a larger force as we go from left to right and that is why the ionic radius decreases along each row.

Now compare ions from different groups which have the same charge or valency, for instance, Li^+, Na^+, ..., Cs^+. The nuclear charge increases as we go from one element to the next and this tends to draw the electrons closer to the centre. However, the *number* of electrons increases and this tends to increase the volume of the ion. Overall, the two effects compensate each other to a large extent and Table 1.1 indeed shows that the radius of ions with the same valency grows relatively slowly as Z increases: from Li^- to Cs^+ the number of electrons increases from 2 to 54 (i.e. 27 times larger) but the radius only increases 2.4 times (and the volume

Table 1.1   Ionic radii in crystals. Values are quoted in nanometres $= 10^{-9}$ m.

Ionic charge

-2	-1	Inert gas	$+1$	$+2$	$+3$	$+4$
		(He) $Z = 2$	Li 0.07	Be 0.035		
O 0.140	F 0.135	(Ne) $Z = 10$	Na 0.100	Mg 0.07	Al 0.05	Si 0.04
S 0.180	Cl 0.181	(Ar) $Z = 18$	K 0.133	Ca 0.100	Sc 0.080	Ti 0.065
Se 0.195	Br 0.195	(Kr) $Z = 36$	Rb 0.150	Sr 0.110	Y 0.100	Zr 0.080
Te 0.220	I 0.215	(Xe) $Z = 54$	Cs 0.170	Ba 0.135	La 0.110	Ce 0.095

Table 1.2   Testing the validity of the concept of ionic radii.

Alkali halide	Distance between +ve and −ve ions measured in the crystal (nm)	Value calculated from the ionic radii of Table 1.1 (nm)
LiF	0.202	0.205
LiI	0.300	0.285
NaF	0.232	0.235
NaCl	0.281	0.281
NaI	0.324	0.315
KF	0.267	0.268
KCl	0.314	0.314
KI	0.353	0.348
RbF	0.282	0.285
RbI	0.367	0.365

by a factor of 14). By contrast, the ionic *mass* is approximately proportional to Z. This means that the average density (mass per unit volume) of the ions, and hence of the solid or liquid formed by those ions, increases with atomic number. For example, LiF has a relative density of 2.64 and CsI of 4.51.

In **metals**, the basic particle is a positive ion, the atom having lost one, two or three electrons from the outer shell. Examples are Na^+, Mg^{2+}, Al^{3+}, etc. (see p. 83 for more details). The detached electrons move freely in the metal and form a negatively-charged 'sea' (called an *electron gas*) in which the positive ions are submerged. This not only guarantees the neutrality of the whole assemblage, but is responsible for the cohesion of the metal as well. The distance between two neighbouring ions in a metal can be measured experimentally, and it is found that the structures of all metals are very well accounted for by a model in which *the metallic atom is also represented by a hard sphere* whose radius is one-half of the experimental distance between nearest-neighbour ions.

Table 1.3 Atomic radii in metals.

Element	Z	Atomic radius (nm)	Element	Z	Atomic radius (nm)	Element	Z	Atomic radius (nm)
Li	3	0.152	Ca	20	0.197	Cs	55	0.266
Na	11	0.186	Cr	24	0.125	Ba	56	0.217
Mg	12	0.160	Fe	26	0.124	Au	79	0.144
Al	13	0.143	Cu	29	0.128	Pb	82	0.175
K	19	0.231	Ag	47	0.144	U	92	0.150

That is the method used to establish the values given for metals in Table 1.3.

A comparison of Tables 1.1 and 1.3 shows that *the radius of a metallic atom is always greater, and sometimes much greater, than the ionic radius of the same chemical element.* Thus, the ionic radius of Na^+ is 0.10 nm while that of the Na metallic atom is 0.186 nm. The difference is in the same direction but much less between the Ag^+ ion (0.126 nm) and the metallic Ag atom (0.144 nm). Since we have just said that the 'particle' of the metal was also an ion, this difference needs explaining. The answer to the problem lies in the nature of the bond. In the ionic bond, the positive ion is attracted to a negative ion. In a metal, the neighbouring ions are both positive and would clearly repel each other if they were on their own. It is only the negative charges of the electron 'sea' which maintain the positive ions at their observed distance apart, but this is still somewhat larger than the interionic distance. The remarkable thing is that, although metals are really quite empty (see Fig. 4.34), the atoms still behave like hard spheres in contact.

Another characteristic of the radii of metallic atoms is that, broadly speaking, they do not vary much with atomic number. The smallest is beryllium ($Z = 4$, 0.111 nm), the largest caesium ($Z = 55$, 0.266 nm). Lithium ($Z = 3$), cadmium ($Z = 48$) and uranium ($Z = 92$) all have metallic radii quite close to each other (about 0.150 nm). It follows that the density of metals increases on average in proportion to the atomic number, and that is why elements of small atomic number are commonly called light elements and those of high number heavy elements. However, as Fig. 1.8 shows, the increase of density with Z is not uniform.

We now turn to the more complex case of arrangements where the basic particle is the atom with covalent bonds as in molecules or in solids of the diamond type. In the case of ions, the very stable outer electron shell remains the same whatever other ion is in contact with it. For *atoms* in interaction with others, only the core remains unchanged, while the state of electrons outside the core is modified by the bond in a way that depends on the nature of the neighbouring atom. Not only that, but covalent bonds have specified directions and are said to be **directed** bonds (we saw an example with carbon). As a result, an atomic model using spheres of fixed radius is too simple and inadequate.

What is it exactly that we determine from experimental measurements on crystals? It is the distance between the centres of

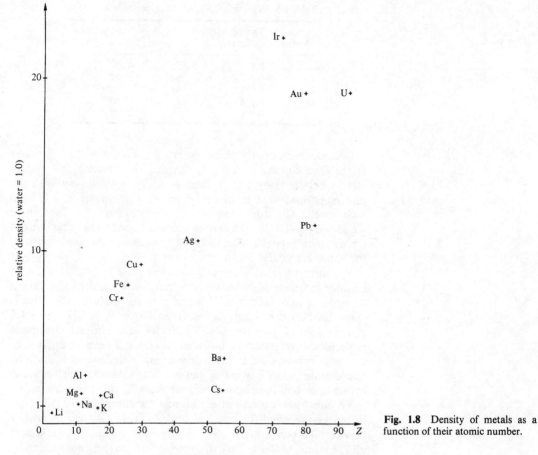

Fig. 1.8 Density of metals as a function of their atomic number.

a pair of neighbouring atoms of *given chemical species* and *linked by a bond of given type*. That is the distance which proves to be the same for all compounds. Thus the C—C distance in diamond is 0.154 nm, which is also the length of the C—C covalent bond in all aliphatic molecules. The length of the double bond C=C is only 0.134 nm and that of the triple bond C≡C is 0.120 nm. There are tables available giving the distances between two given atoms linked by a covalent bond and several examples are to be found in Table 1.4.

Table 1.4 Distances between atoms linked by covalent bonds.

Bond	Distance (nm)	Bond	Distance (nm)
C—C	0.154	N—N	0.146
C—C	0.1395	N—H	0.102
(in benzene)		N—C	0.147
C=C	0.134	O=O	0.121
C≡C	0.120	(O_2)	
C—H	0.108	O—H	0.097
C—Cl	0.177	O—C	0.143
N≡N	0.110	Cl—Cl	0.199
(N_2)			

Table 1.5   Van der Waals radii of several atoms.

Element	Van der Waals radius (nm)	Element	Van der Waals radius (nm)
H	0.117	I	0.210
C	0.180	Ne	0.160
N	0.157	Ar	0.192
O	0.136	Kr	0.201
Cl	0.178	Xe	0.221

Four covalent bonds emerge from a carbon atom in the direction of the four corners of a regular tetrahedron, making an angle of 109.5° between any two of them (see Fig. 1.4(c)). Now we have just mentioned that in diamond the C—C distance is 0.154 nm with each C atom linked to four others. In the chain —CH_2—CH_2—CH_2—, the carbon atom is bound to only *two* Cs but the distance between them is still 0.154 nm and the angle between the bonds is still 109.5°.

In contrast to that, if the neighbouring atom to carbon is that of *another* molecule (to which it is therefore not bonded), the closest distance of approach is as if it had a radius of 0.180 nm. This is called its *Van der Waals radius* (see Chapter 4, p. 75). Table 1.5 gives values of Van der Waals radii for atoms found in organic molecules. Inert gases are included here as well because the lack of bonding between their 'monatomic molecules' means that the appropriate radius is also the Van der Waals radius, which is found in fact to be half the distance between atoms in these substances.

To construct a model of a molecule, its atoms are placed at the distances predicted for covalent bonds (Table 1.4) and the angles between the bonds are fixed at their proper values. The centres of all the atoms at the *surface* of the molecule are then surrounded by their Van der Waals spheres: the envelope of these spheres defines the external shape of the molecule. In the interior of the molecule the different Van der Waals spheres overlap, but that is not serious since the atoms of the same molecule are linked to each other anyway. It is important not to confuse the spherical models of metallic atoms or of ions, which are *impenetrable*, with the Van der Waals spheres. Figure 1.9 shows two examples of molecular models: those of benzene, C_6H_6, and of a complex molecule designated by the abbreviation TBBA.

In condensed states of matter, the contact between molecules is determined by their external shape, and models such as those of Fig. 1.9 enable us to predict how they will pack together. That is just one example of the use physicists make of such models. We have come a long way from the abstract idea of matter as composed of a collection of atoms too small to be directly accessible to our senses and having properties completely different from those of the material objects around us. We have arrived at the idea that a useful model of the structure of the material can be built by assembling hard balls, or objects of more complex shape, and observing a set of rules. One such rule which is always obeyed, and is very simple because it conforms to our everyday experience of visible objects, is that *an atom can only be placed at a position where other atoms leave sufficient room.* This is one of the fundamental ideas used in building structures of matter, particularly of molecules: it is called 'steric hindrance' by chemists.

It is true that some of our explanations have been a little over-simplified—atoms *can* be slightly compressed or deformed, and there are certainly some particularly complicated cases—but the basic idea, which is of surprising generality and simplicity, is not altered and we must take full advantage of it.

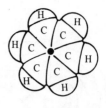

(a)

The hardness of atoms: Pauli's principle in quantum mechanics

All the above discussion depends on the fact that as one atom approaches another there is a minimum distance between them which cannot be reduced: 'something' prevents them from penetrating one another. Classical physics offers no explanation of this. Bohr's atom, with its point-like electrons following paths around the nucleus, is more or less as empty as the solar system, so why can't the electron trajectories overlap as the nuclei approach each other?

The answer is that *classical mechanics has to be abandoned in favour of quantum mechanics.* Although we find that quantum theory is not essential for the great majority of topics dealt with in this book, some of its results are indispensable. We shall make use of such results but shall accept them without justification—they are ideas simply 'taken over' so that progress can be made without dragging in the general principles of the theory.

An electron in a given atom must occupy one of a set of possible 'states', each one distinct from the others and labelled by a code made up of several quantum numbers, integral or half-integral. When the code is understood and the labels known, some of the observable properties of the electron can be calculated—for example, its binding energy with respect to the nucleus.

Pauli's principle asserts that each state in an atom cannot be occupied by more than one electron. Thus, there are only two states with quantum number $n = 1$, those with the maximum energy. It follows that there can be no more than two electrons in this shell, the 1s or K shell. Similarly, there can be no more than eight electrons in the L shell for which $n = 2$, and so on. If an atom has more than ten electrons in all, then the third shell *must* be occupied. Pauli's principle completely explains Mendeleev's periodic classification of the chemical elements.

Consider first two *ions,* one +ve and one −ve, placed near to each other and pushed together. If one were Na^+ and the other F^-, both would have completely filled K and L shells and neither could accept an electron from the other. The only thing that can happen is that the electrons remain with their respective ions, which must then be sufficiently far apart for neither to disturb the other. The effect of the conditions imposed by Pauli's principle manifests itself as a repulsion between the ions. This balances the electrostatic attraction and fixes an equilibrium distance apart. To a good approximation,

(b)

Fig. 1.9 Molecular models. (a) Benzene molecule, C_6H_6. (b) Molecule of TBBA, terephthal-bis-butyl-aniline, with the formula $C_4H_9-\phi-N=CH-\phi-CH=N-\phi-C_4H_9$, where ϕ stands for the phenyl radical C_6H_4.

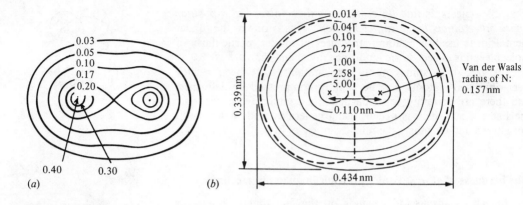

Fig. 1.10   (a) Map of the electron density distribution in the meridian plane of the H_2 molecule. At the centre is a col between the two peaks centred on the hydrogen nuclei. (b) N_2 molecule: meridian sections through the surfaces of equal electron density with values in 10^3 electrons per nm^3. The nuclei are 0.110 nm apart (Table 1.4). The broken curve is the envelope of the Van der Waals spheres of the two N atoms. Nearly all the total charge on the electrons is inside this envelope.

the process can be represented as the attraction of hard spheres up to the point where contact is made.

To explain the *covalent* bond between two atoms, we take the simplest example, starting with two H atoms separated from each other and each with its electron. The 1s shells are not complete, since a second electron can occupy the second available 1s state in each atom. As the nuclei approach each other, the situation changes: the 1s states become deformed by the electric field created by the two protons, and they are replaced by an *'orbital' encircling the two nuclei* which *both* electrons can occupy. The cohesive energy of this molecule is greater than that of the two separate H atoms. The excess, which is the binding energy, passes through a maximum when the distance apart of the protons is 0.074 nm and decreases rapidly when the nuclei approach each other more closely than that. Since a system always tends spontaneously towards the state of greatest cohesive energy, the atoms attract each other when they are more than 0.074 nm apart and repel when they are closer: the H—H covalent bond has been formed.

That example can be rigorously treated and it shows exactly how and why two electrons can share one orbital. The negative charge density is distributed throughout the molecule as indicated in the 'map' of Fig. 1.10(a). What is remarkable about this is that a small part of the charge is found concentrated at the centre of the line joining the two nuclei. It can thus contribute to the bond by attracting the two positive nuclei. That is what is being suggested diagrammatically but very crudely by the bar placed between two H atoms thus: H—H, or by a pair of electrons placed between them thus: H:H. In fact, quantum mechanics shows that only about 0.2 of an electronic charge is situated between the nuclei—quite a small proportion of the total negative charge. The distance between the protons in the H_2 molecule (0.074 nm) and its dissociation energy (4.747 eV or 458 kJ $mole^{-1}$) can be calculated exactly and agree perfectly with experimental values.

In more complex atoms, the cores with their complete shells cannot penetrate each other, and it is the outer electrons which form one or more pairs occupying orbitals and producing the valence bonds. As in H_2, the distance apart of the nuclei is well defined (see Fig. 1.10(b) for N_2).

So we find, then, that the behaviour of electrons obeying quantum mechanical principles can be represented on a human scale by a material model involving contact between hard objects. Such a model is clearly not a description of the actual system of electrons, etc., but it is something that enables us to understand and predict atomic phenomena rather more easily.

Furthermore, we accept the postulate we called Pauli's principle because of its consequences—such as, for example, the fact that atoms are impenetrable and it follows that solid matter is too. As Feynman says in his *Lectures on Physics*, every time we lean on the table and our hand encounters resistance, we are verifying Pauli's principle.

There is too great a tendency to imagine that everyday occurrences are a matter for classical and familiar theories, while discoveries associated with the large and complex apparatus of modern physics can only be explained by ideas that are of recent origin, very complicated and even contrary to common sense. The example we have just quoted shows that this is wrong. A baby playing with its toys understands the impenetrability of matter, yet we need quantum mechanics and the mysterious Pauli principle to explain it.

Comment on the absolute value of the densities of solids

When comparing the atomic radii of different elements, we accounted for the variation of the density of solids as a function of the atomic number of the element. Let us now consider instead the *absolute* value of the densities, which determine how many atoms there are in a given volume. For all substances, these densities are of the order of a few g cm^{-3} (from a few tenths to about twenty), figures that are related to the absolute values of atomic and ionic radii because condensed states of matter can be very crudely represented by spherical atoms or ions packed closely together.

Now the atomic or ionic radii of all the chemical elements are given by the Bohr radius r_0 multiplied by a numerical factor between about one and four. The radius r_0 in the Bohr theory represents the radius of the smallest circular electron orbit in atomic hydrogen; in quantum theory, r_0 is equal to the most probable distance of the electron from the nucleus of the hydrogen atom in its ground state (Fig. 1.7(a)). A quantum mechanical derivation gives the following expression for r_0:

$$r_0 = \varepsilon_0 h^2 / \pi m_e e^2$$

where ε_0 is the permittivity of free space, introduced because the formula is given in SI units. Numerically, $r_0 = 0.0529$ nm.

The value of r_0 thus depends on three fundamental constants of atomic physics: e, the elementary electronic charge, m_e, the mass of the electron, and h, Planck's constant, which determines the size of quanta in our universe[1].

[1] $h\nu$ is the elementary packet of energy for an oscillator of frequency ν. Planck's constant is also what ultimately limits the precision of our knowledge about atomic systems. No matter how perfect the method of measurement, it gives the

The radius of the hydrogen atom, and hence of all atoms, is fixed by these three fundamental constants or, more exactly, by the combination $h^2/m_e e^2$. That is the quantity which determines the *volume* occupied by a given number of atoms. However, the *mass* of an atom is fixed by that of the proton and in going from hydrogen to uranium it varies from one to about 237 proton masses. The density of solids is thus related to the mass of the proton together with e, m_e and h, so that as familiar a quantity as the macroscopic size of an object of given mass is linked directly to the fundamental constants of physics. The mass of a pebble of given volume would be a million times greater or smaller if the electronic charge were ten times greater or smaller than it actually is in our universe.

---

product of uncertainties in the simultaneous position and momentum of a particle. In quantum mechanics, these two quantities are so related that the more accurately one of them is measured, the less accurately can the other be ascertained.

2 The two states of matter: order and disorder

Many readers will no doubt have learnt in previous science courses that a given substance can, depending on the temperature and pressure, exist in one of three states: solid, liquid and gas. These states are distinguished from each other by differences in macroscopic properties which are so well known that there is little point in recalling them here (for instance, differences in density, viscosity or elasticity).

If, instead of these macroscopic properties, we use as our criteria the characteristics of atomic models of the substance, we are led to a different classification of its physical states. For we then find that there are only **two** states separated by *one* clearcut line of demarcation: one is known as the *disordered* state and the other as the *ordered* one. Of course, at some stage we must show how this new classification is related to that of the three states with which we are more familiar.

So as to make our explanation quite clear, we shall consider in this chapter only those substances in which the basic particles are molecules that remain unchanged in all forms of the material. (The simplest cases are the inert gases consisting of monatomic molecules, but small molecules like H_2O, C_6H_6, CO_2, etc., can be used just as well.) Thus, the only thing that will change as we go from one state to another will be the *mutual arrangement* of the molecules.

The disordered state

The perfect gas

First of all, we consider matter in the form of a gas or, more precisely, a gas at a low or very low pressure (1 atmosphere or less). Under these conditions we know that the material *is very dilute or dispersed*. We can illustrate this as follows, using an example in which some orders of magnitude can be estimated.

Take the inert gas neon under ordinary conditions of pressure and temperature (1 atmosphere or 1.013×10^5 Pa, and 0°C or 273 K. Note that in this chapter and the next we shall often use the *bar* as the unit of pressure which is *exactly* 10^5 Pa and *approximately* 1 atmosphere). The volume of 1 mole is then 0.0224 m³, a volume which thus contains 6.02×10^{23} atoms. Now Avogadro's law states that in a given volume of *any* gas at the same temperature and pressure, the number of molecules is independent of their nature—it is the same for all 'perfect' or 'ideal' gases and does not depart much from that number in real gases. In other words, under normal conditions a molecule of any gas whatsoever has at its disposal a volume equal, or approximately equal, to $(0.0224 \text{ m}^3)/(6.02 \times 10^{23})$. *This is the volume of a cube whose side is 3.3 nm.*

At the same time, we remember that atoms and small molecules themselves have diameters of *only a few tenths of a nanometre.* This means that the model of a gas at low pressure consists of particles which are separated by distances large or very large compared with their own sizes. *There is thus practically no interaction at all between molecules*: when they collide with each other, they behave like the 'material points' of classical mechanics. So, for a gas of low density[1], the nature and internal structure of the molecules is irrelevant: it is only their *number per unit volume* which determines the pressure. That is the meaning of Avogadro's law. (At a later stage, on p. 44, we shall have to distinguish between monatomic and polyatomic molecules for some properties of the gas, but for the moment we can ignore such a distinction.)

The absence of interaction means that molecules have no influence on each other, and that the position of one of them is independent of the positions of all the others. They are thus in a state of total disorder. It is important to realize that the last statement is not *quite* exact, for when $(N-1)$ molecules are already in their places, a certain volume is in fact forbidden to the centre of the next one, the Nth. This is because it must be outside the sphere of influence of those $(N-1)$ others: atoms, remember, are impenetrable. However, this restriction is very small. For example, the atomic radius of neon (Table 1.5) is such that the atoms themselves in normal conditions occupy no more than one-thousandth of the total volume, a negligible proportion. In any case, disorder tends to become *complete* as the density of the gas decreases more and more. This, of course, is precisely when it is becoming more and more like the state we call that of the perfect gas.

The **perfect gas,** *therefore, corresponds to a model with molecules in complete disorder and with no interaction between them.*

Thermal motion

At this point we must introduce a new feature into our structural models, that of thermal motion. This is a universal phenomenon—that is, it occurs in every state of a material—but it takes a particularly simple form in a perfect gas, as we shall see shortly.

Firstly, we should realize that *models of structures are in general never static ones.* The basic particles do not remain stationary but instead perform a perpetual motion which, at a given temperature, can be neither speeded up nor slowed down. The energy of a particle due to its movement depends solely on the temperature, being zero[2] at absolute zero ($-273\ °C$) and increasing as the temperature rises. We shall see, in fact, when we study perfect

---

[1] The term 'density' of a gas often denotes the *number of molecules* per unit volume rather than the *total mass* per unit volume.
[2] more or less. In fact, a very small residual motion does persist at absolute zero, something that is predicted by quantum mechanics.

gases in more detail, that the energy of thermal motion is used as a measure of the temperature.

Without thermal motion, our model of a gas would make no sense since the molecules would fall to the bottom of any vessel under the action of their own weight. The speed of molecules at ordinary temperatures is of the order of several hundred metres per second, so that in spite of their weight the trajectories are almost straight lines. A simple calculation shows that their fall under gravity at these speeds is no more than about 1 μm in a distance of 10 cm. The movement of the molecules causes them to collide with other molecules and with the walls of the containing vessel, producing a pressure. These collisions make them deviate from their straight paths so that the resultant trajectories are very complicated patterns of broken lines. When all the molecules are taken together, such motion produces the complete disorder in their positions that we have already noted.

We said above that the molecule of a perfect gas behaves like a 'material point' of classical mechanics. The motion of such a molecule at a given temperature can be specified by a single quantity whatever the gas—that is, its mean kinetic energy of translation $E_m = \frac{1}{2}mv^2$. However, E_m depends only on the temperature, whereas the mass m and the mean speed v both depend on the type of molecule. We shall see in Chapter 3 how such a model is used to make calculations in the kinetic theory of gases.

Real gases

Starting from the simple case of a perfect gas at very low density, we progressively increase the pressure while keeping the temperature constant. The volume decreases, and so the number of molecules per unit volume gets larger. It follows that the average distance between them gets smaller.

For neon gas at 0 °C, the volume available to one molecule at 10 atmospheres pressure is $(1.5 \text{ nm})^3$, while at 100 atmospheres it is only $(0.7 \text{ nm})^3$ compared with $(3.3 \text{ nm})^3$ at 1 atmosphere. At the higher pressures, a model using disordered molecules is still valid but the disorder is no longer complete and without any restriction. For one thing, the volume forbidden to the molecules is no longer negligible; for another, a molecule in the course of its travels will be close to other molecules for an appreciable proportion of the time. Hence we can no longer neglect the interaction between molecules and we are into the domain of what we call a *real gas*. Gases under high pressure fall into this category and for some the pressure need only be 1 atmosphere or so. In these cases, the gases are a long way from the ideal state and not just because of the number of molecules per unit volume. It is also due to the strength of the interactions between them because of their size and because of certain other particular properties (see, for example, H_2O in Chapter 7). So, for the same molar volumes at the same temperature, different gases can be quite a long way from being 'perfect'.

The laws followed by perfect gases can be rigorously derived from the model of complete disorder, as we shall see in Chapter 3.

Real gases are only an approximation to this model and so do not exactly follow the perfect gas laws. The deviations become greater as we get further away from the perfect state—that is, as the density of a particular gas becomes too high.

From compressed gases to liquids

What happens when, under increasing pressure, a density is reached which is much greater than that of gases under normal conditions? We shall follow the process using a particular example so that we can be quite precise about all the parameters.

We take a mole of the substance H_2O—that is, we take 6.02×10^{23} molecules of H_2O with a total mass of 18 g. It is contained in a vessel in which the *temperature* and *pressure* are fixed from outside. Once these are fixed, the *molar volume* has a value at equilibrium which can be looked up in tables of the physical properties of water. However, instead of molar volume, we shall use either v, the *volume per molecule*, or the ratio v/v_0, where v_0 is the actual volume of the H_2O molecule (in fact around 0.0168 nm^3 for the structure given in Fig. 7.6).

We start with H_2O in the gaseous state at point A in Fig. 2.1, where the parameters are fixed at $p = 0.02$ bar (or 2000 Pa) and

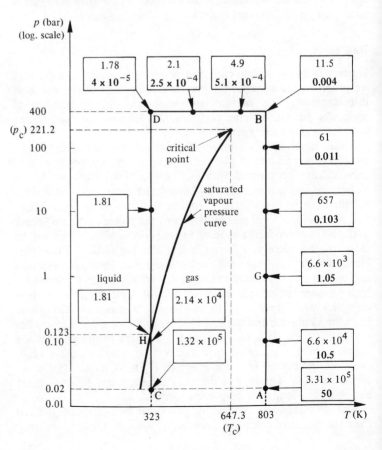

Fig. 2.1 Properties of fluid H_2O as a function of its state (temperature T and pressure p). The upper figures in the rectangles give the ratio of v, the volume available per molecule, to v_0, the volume of the H_2O molecule itself (0.0168 nm^3). The lower figures, in bold type, give the isothermal compressibility $-(1/v)(dv/dp)$ in bar^{-1} (equal to $1/p$ for a perfect gas only). The curve is that of the saturated vapour pressure and represents the boundary between the liquid and gaseous states; it terminates at the critical point.

$T = 530\,°C$ (803 K). These conditions fix the value of v/v_0 at 3.3×10^5. The material in this state is a typical 'real gas' of water molecules and is often called 'water vapour', although there is no precise significance attached to the word *vapour* as distinct from the word *gas*.

We now compress the gas to a pressure $p = 400$ bar (395 atmospheres) keeping the temperature constant at 803 K so that we move vertically up to state B in the figure. During this compression, the volume diminishes regularly and monotonically from an initial value of v/v_0 of 3.3×10^5 down to a mere 11.6. The substance in the vessel remains perfectly homogeneous throughout and *no sharp transformation is observed*.

Let us now do another experiment, starting instead from state C where $p = 0.02$ bar again but this time $T = 50\,°C$ (323 K). Here v/v_0 is 1.3×10^5 and the material is also in a gaseous state. However, when it is compressed at constant temperature as in the first experiment, things do not proceed in quite the same way. At a pressure of $p = 0.123$ bar (state H), where v/v_0 is 2.14×10^4, there is a *definite discontinuity*. When the volume available to the material is decreased at the same pressure, the gas *liquefies*, and remains at H. Up to point H, the substance was homogeneous, but now it divides into two phases with different physical properties. One phase is the gas that still persists and the other is a *dense* phase with v/v_0 only 1.81: the *liquid*, which collects in the bottom of the vessel and which is bounded by an upper free horizontal surface underneath the gaseous phase. While the volume is decreasing, the pressure remains constant and the material remains at point H at the same time as the liquefied portion increases. When *all* the substance has liquefied, the pressure can again increase and the volume can also decrease, but only by a very small amount: v/v_0 is still 1.78 when the pressure has become as high as $p = 400$ bar at point D. In this second experiment, then, we first of all created a liquid phase and then compressed it from $p = 0.12$ bar to $p = 400$ bar.

Now in state B ($p = 400$ bar, $T = 803$ K) we had a compressed gas at the end of the first experiment. Suppose we now take this and lower the temperature from 803 K to 323 K at constant pressure. We move from state B to state D and observe *no discontinuity* in the physical properties during the change: there is merely a progressive decrease in v. Naturally we should be inclined to say that in state D we then had a compressed gas at $p = 400$ bar and $T = 323$ K—but this same state is obviously what we called a liquid in the second experiment when we reached D by a transition from C.

These experiments, although very classical and well-known, are quite fundamental and we must ask what conclusion we can draw from them now that we have followed the transitions in great detail. The important conclusion is that *it is possible to pass in a continuous fashion from the gaseous to the liquid state* (for instance, via the path A → B → D) with no discontinuity whatsoever. The model of the structure, although modified, must remain *essentially* the same, and hence *the liquid, like the gas, has a disordered structure.* The difference between gas and liquid arises from the change in density. Along the transformation path from A → B → D, the fluid remains homogeneous, but the value of v/v_0 decreases from 3.3×10^5 at A to 11.6 at B to 1.78 at D.

In the initial state, A, the fluid is a gas in which the distance between molecules is of the order of 120 diameters. This makes it approximate quite closely to a perfect gas. In state B, the molecules are on average quite near to each other at distances of around four diameters. Finally, in state D, the distances between neighbouring molecules have become of the same order as the molecular diameters themselves: they are packed quite closely together and many are in contact.

The liquid state can thus be represented by a disordered collection of molecules pressing one against the other, rather like tiny marbles piled up in the bottom of a bag. Between the liquid and the gas there exists a continuous series of intermediate states with progressively greater numbers of molecules per unit volume. We are therefore quite justified in allocating the liquid and gas to the same class of *disordered structures*.

The transformation from gas to liquid by condensation and its inverse: vaporization

In the above discussion we have concentrated on the *continuous* transition between gas and liquid, such as that by the path A → B → D in Fig. 2.1. However, there is also the possibility of a *discontinuous* transition, which can occur within a limited range of temperatures. Referring to Fig. 2.1, suppose that we carry out a whole series of compressions similar to those from A to B and from C to D, but at various temperatures ranging between 323 K and 803 K. We should find that there would be a discontinuity similar to that at point H, but only for temperatures *below* 647.3 K. This is known as the critical temperature T_c, and above it *there is always a continuous transition.*

Condensation occurs abruptly at a pressure called the *saturated vapour pressure*, which increases as the temperature rises, following a curve that terminates at the critical point ($p_c = 221.2$ bar, $T_c = 647.3$ K in Fig. 2.1). The abrupt change in volume as condensation occurs is very great at lower temperatures, but gets smaller as the critical point is approached. At the critical point itself, the densities of liquid and gas become equal, as is shown in Table 2.1.

We saw, along the path A → B → D, that the molecules became progressively and continually closer, and that it was the same from C to H. However, at the particular state where condensation occurs, at H itself for instance, the molecules do not become closer in a continuous way if the volume is now made smaller. On the contrary, small isolated regions or islands are formed where the molecules pack more tightly (Fig. 2.2), but leave the remainder of the molecules still dispersed with the same density as before, thus keeping the pressure constant. These small islands collect together

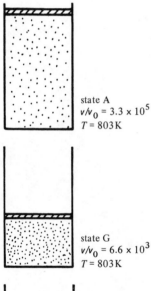

state A
$v/v_0 = 3.3 \times 10^5$
$T = 803$ K

state G
$v/v_0 = 6.6 \times 10^3$
$T = 803$ K

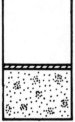

state H
(condensation)
$v/v_0 = 2.14 \times 10^4$
$T = 323$ K

Fig. 2.2 Schematic representation of molecular models (not to scale): the upper figures represent homogeneous fluids at states A and G in Fig. 2.1; the bottom figure indicates what happens during condensation at state H.

Table 2.1 The available volume per molecule for H_2O in the liquid and gaseous states as a function of temperature from the triple point to the critical point.

Temperature		Saturated vapour pressure	Ratio of the volume per molecule, v, to the actual volume of the H_2O molecule, v_0	
(°C)	(K)	(bar $= 10^5$ Pa)	Liquid	Vapour
0.01	273.16	0.006	1.78	3.68×10^5
20	293	0.023	1.79	1.03×10^5
50	323	0.123	1.81	2.15×10^4
100	373	1.013	1.86	2.99×10^3
150	423	4.76	1.95	7.02×10^2
200	473	15.5	2.06	2.27×10^2
250	523	39.8	2.24	89.5
300	573	85.9	2.51	38.7
350	623	165.3	3.11	15.7
374.15	647.30	221.2	5.66	5.66

and grow into droplets[1], so that the volume can become smaller while the gas still preserves the same pressure. It is only when *all* the fluid has condensed into a liquid that the pressure can increase beyond the value of the vapour pressure.

The molecules have a spontaneous tendency to form small clusters of greater density because of the Van der Waals attraction between them (see p. 75). The condensation lowers the energy of the system and some heat is released. However, the system as a whole is *less disordered* when the small denser clusters are present than when it was homogeneous, with the molecules distributed completely at random. Now it is a consequence of the second law of thermodynamics that a system tends to change spontaneously towards the most disordered possible state. We thus have two contradictory effects[2]. They exactly compensate each other when the system is in equilibrium, as it is at any point on the curve of saturated vapour pressure against temperature.

We thus have a qualitative explanation of condensation. Can we go any further? The phenomena we have been describing are certainly *compatible* with our models of a gas and a liquid but we ought to understand that, so far, the models do not allow us to make *predictions*. To emphasize this point, imagine that we have the two phases existing together, both disordered: the gas, less dense, with its molecules in rapid motion, and the liquid, more dense, with its molecules also moving but remaining closely packed against each other. In the course of its motion, it may happen that a gas molecule will collide with the liquid and remain stuck to it. Conversely, a molecule of the liquid might escape into the gas. When we say that the two phases are in equilibrium, we are simply

[1] In the presence of a gravitational field, the droplets of the liquid phase fall to the bottom of the container, but if gravity is ineffective the droplets and the gas together form a stable mist.

[2] Remember that the stability of a system is conditioned by the relative influences on the free energy F of the internal energy U and the entropy S because $F = U - TS$.

saying that in a given period of time the same number of molecules would be evaporating and condensing. If the vapour pressure were to be increased, there would be more collisions with the liquid surface and so an excess of condensation over evaporation. The quantity of liquid would increase until the equilibrium was re-established at the new pressure. Conversely, if the pressure were lowered, there would be an excess of evaporation. Thus we can understand the *idea* of an equilibrium pressure, but we have no way of explaining why it only exists below a certain temperature T_c, nor do we know how to calculate the value of T_c. To do that, we should have to take into account the interactions between molecules which we do not know with sufficient precision.

Thus, we are baulked by a phenomenon as commonplace as vaporization. However, although this shows that we must not ask too much of simple models and must be careful to realize their limitations, we must not lose sight of the fact that, simple as they are, they render an immense service in increasing our understanding.

The compressibility of fluids

Confirmation of the model of a liquid and of its continuity with a gas is provided by values of compressibility under various temperatures and pressures. Compressibility is measured by the ratio of the fractional decrease in volume $(-dv/v)$ to the increase in pressure (dp) producing it—in mathematical terms, it is $-(1/v)dv/dp$. Some values for H_2O are indicated in Fig. 2.1. A gas is highly compressible because of the large spaces between its molecules. However, the compressibility of a liquid is very small because so many of the molecules are in contact and thus are at the minimum distance apart. For that reason, even an enormously great pressure will only reduce the volume by a small amount. Moreover, the reason why a liquid can transmit pressure from one point to another so well is that the molecules press hard on their neighbours, rather like a compact collection of hard balls (Fig. 2.3).

Referring back to Fig. 2.1, the fluid in states A and C is certainly a gas, while that in state D is equally certainly a liquid; between these extremes the compressibility varies continuously. In state B, and in the states between B and D, the molecules are certainly much closer than in a perfect gas, but nevertheless they are not in contact all that often. So do these states represent a highly compressed gas or a liquid of very low density? In fact it is purely arbitrary: either description is valid.

Liquids open to the atmosphere

A liquid exposed to the open air is also in equilibrium with its saturated vapour at a pressure corresponding to the temperature. However, although the presence of air does not affect the equilibrium state itself, it does have a great influence on the way the system reaches that equilibrium. Indeed, if the molecules escaping from the liquid surface can move without restriction into the atmosphere, the vapour in contact with the liquid will never reach

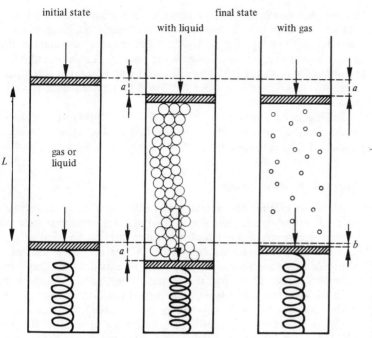

Fig. 2.3 Transmission of pressure by a liquid. Pushing the upper piston down through a distance *a* requires the same force as is needed to compress the spring directly by the same amount. If the cylinder is filled with gas, the force needed to push the upper piston down by *a* is much less and the spring is compressed by only a very small amount *b*.

the equilibrium pressure. In that case, the liquid will evaporate completely. In fact, the speed of diffusion of the molecules through air is limited, so that they form a 'cushion' of vapour at the surface of the liquid at the required pressure. Only enough evaporation takes place to make up for the loss of vapour molecules by diffusion into the air.

If the vapour pressure is low, there is only a small amount of material in the vapour coming from the liquid. To all intents and purposes, the liquid on its own can then be considered as stable. For example, at a room temperature of 20 °C, mercury has a vapour pressure of 0.0016 mbar and silicone oil one of less than one-millionth of a bar. In these cases, the loss by evaporation, even after a whole day, is negligible. Nevertheless, the vapour, dilute as it is, can still have noticeable effects—for example, the smell from certain liquids and the harmful action of mercury vapour on the body.

The vapour pressure of water at 20 °C is distinctly greater with a value of 23 mbar, but the normal atmosphere is not very dry and already contains water vapour at a pressure between 50% and 70% of its maximum value. This reduces the amount of water that needs to evaporate in order to attain equilibrium. So an open vessel containing water to a depth of a few centimetres is apparently stable for some hours. Of course, if the layer of vapour in contact with the surface is continually removed by an air current, the protective cushion is destroyed and, as everybody knows, evaporation is much faster.

Finally, those liquids with high vapour pressures—for example, alcohol (58 mbar) and ether (572 mbar)—are very volatile. That means they do not survive for very long in a vessel open to the air, but must be kept in a confined space, for example in a hermetically-sealed flask. The liquid in such a flask becomes stable when the air above its surface is saturated with the vapour. Air saturated with alcohol contains only 0.12 g per litre and with ether only 1.9 g per litre, so the amount lost from the liquid in a sealed flask is very small.

Boiling occurs when the saturated vapour pressure becomes equal to the atmospheric pressure above the liquid. This is discussed further in Chapter 7.

Thermal motion in liquids

As a result of the inevitable thermal motion, a static molecular model is no more adequate for a liquid than for any other state of matter. However, since liquid molecules are packed so tightly together, their movements cannot be as free as those in a gas. Many of them are so confined by their neighbours that they can only vibrate around a centre that is more or less fixed. As soon as a molecule wanders away from its preferred position among all the others, collisions from its neighbours tend to take it back again. It performs a sort of vibration with a frequency of around 10^{12} to 10^{13} Hz, a similar range of values to the frequencies of the atomic vibrations in all other condensed states of matter.

If the movements of a number of neighbours happen by some chance to act in concert, then a molecule may be made to exchange its position with another one. Since it is then displaced from its original position, this gives rise to **diffusion**. For example, two miscible liquids such as alcohol and water will slowly mix together even if they are not disturbed or shaken in any way. It has been established that at 20 °C a molecule of water will have moved on average a distance of 0.04 mm from its original position after one second, and will have moved as much as 0.4 mm after 100 seconds. At 50 °C, the distances are respectively 0.08 mm and 0.8 mm. By way of comparison, at 20 °C a molecule of air moves on average 6 mm in one second, a fact which clearly demonstrates the difference between the freedom of molecular movement in a gas and that in a liquid.

On the same scale, another piece of evidence for thermal motion of molecules in a liquid is provided by **Brownian motion**. A particle immersed in a liquid is always being buffeted by the molecules of the liquid. Given the random nature of the collisions, their effects will exactly balance out if a sufficiently large number occur simultaneously. However, if the particle is very small (<1 μm), there are not very many near neighbours—few enough for exact compensation *not* to occur at every instant. As it dances around in an irregular manner under the action of the random collisions, such a particle can in fact be observed under a microscope (Fig. 2.4)[1].

[1] Dust particles observed in a beam of light passing through air also undergo perpetual and irregular movements. However, this is not due to Brownian motion, but is the result of small air currents produced by inequalities in temperature and pressure.

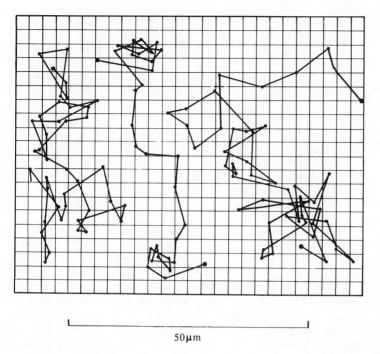

50 μm

Fig. 2.4 Brownian motion of a particle of diameter 0.5 μm suspended in water. During three successive experiments, the horizontal projection of the positions of a given particle is marked every 30 seconds. (Experiments of J. Perrin, *Les Atomes*, Presses Universitaires de France, 1970.)

This Brownian motion has three features to be noted. Firstly, we cannot influence it in any way since it is the result of thermal movements. Secondly, since it can occur in an isolated system with an absolutely uniform and constant temperature, it requires no input of energy from outside. Thirdly, however, it cannot be used for doing work because it is completely disordered.

The viscosity of liquids

The *fluidity* of a liquid is also explained by the model consisting of closely-packed molecules behaving like tiny billiard balls. When it is not resting in the bottom of a vessel in equilibrium, a liquid flows—that is, it deforms and moves until a new equilibrium is established. An example is illustrated in Fig. 2.5(b) where an oil is shown being poured from one vessel into a lower one. The same thing happens with fine dry sand as shown in Fig. 2.5(a), and here we can see that the flow occurs by the rolling and sliding of sand grains over each other. In the same way, in the liquid, *the molecules slide over each other* as well. Certainly there are internal cohesive forces which keep the liquid in a compact form with a constant *volume*, but these forces are not sufficient to prevent the change of *shape* that occurs quite readily. If the molecular sliding is easy, the liquid is very fluid and we say that its *viscosity* is low. That is the case with water and even more so with ether (Table 2.2). In contrast to those examples, oil flows quite slowly and is said to be *viscous*. Thick oil poured into a vessel does not achieve its equilibrium state with a horizontal upper surface until a certain time has elapsed that may be between a few seconds and a few minutes.

(a)

(b)

Fig. 2.5 (a) Flow of sand grains in an egg-timer. The free surfaces of the sand in the two parts of the vessel are not horizontal, but they would become so if the grains were made mobile by shaking very lightly. (b) Flow of a viscous liquid.

Table 2.2 Viscosity of some liquids compared with that of water.

Liquid	Temperature (°C)	Coefficient of viscosity $(10^{-3}$ Pa s$)$[1]
Ether	20	0.23
Benzene	20	0.64
Cyclohexane	20	0.98
Water	20	1.00
Glycerine	20	1.41×10^3
Machine oil		
light	11	1.15×10^2
heavy	11	6.6×10^2
Cane sugar	125	1.0×10^5
Fused silica	$\begin{cases} 2200 \\ 2500 \end{cases}$	7.0×10^5 4.6×10^4

[1] formerly called centipoise.

Viscosity is a property that can be measured precisely and expressed as a coefficient whose value varies considerably from one liquid to another (see Table 2.2). For any given liquid, the viscosity increases continuously as the temperature falls. So, for instance, a light oil that is not very viscous at high temperatures becomes a thick oil at room temperature. If it is further cooled to below 0 °C its consistency thickens and takes on the character of a soft grease or wax, while below about −20 °C to −30 °C it becomes 'pitch', similar to a soft solid, with its own shape but very easily deformed.

Ordinary glass exhibits an even greater transformation. At a very high temperature (800 °C) it is a very viscous liquid. On cooling, it passes through a soft pasty state until at ordinary temperatures it appears to be a very hard solid, not capable of being deformed.

Amorphous solids

This is the name given to solids produced by transformations such as those we have just described. The change from the liquid to the solid state as the temperature falls takes place without any discontinuity in physical properties. This implies that *the change is accompanied by no essential changes in the atomic structure.* We also notice that there is only a very slight variation in the density during such transformations.

We conclude from this that the *arrangement* of the molecules remains the same as in liquids: a disordered packing with very similar densities. The difference is that, while in liquids the molecules slide easily over each other, in pastes and waxes it is more difficult and in glasses impossible.

We have thus brought within our class of disordered structures some materials, such as glass, that are commonly thought of as solids: they are called **amorphous or glassy solids** and are obtained from the liquid state by the continuous transformation we have described ('amorphous' means without form or structure, i.e.

disordered). Conversely, when heated, the amorphous solid shows a gradual softening into a paste-like state before becoming a viscous liquid: there is no sharp melting point. Such materials are not very numerous, at any rate in comparison with those which exist as **true solids**, as we shall call them in the next section. Amorphous solids are more like liquids with the disordered structure frozen in.

In these solids, as in all other materials, the molecules are undergoing thermal motion, but here it takes the form of localized vibrations about a fixed centre. It is thus very difficult for a molecule to exchange its position with a neighbour. It follows that diffusion (p. 28) is far slower in amorphous solids than in liquids by a factor of the order of 1000. Moreover, because the molecules are only displaced with some difficulty, the whole structure has a rigidity more characteristic of the solid state. An amorphous solid has a shape of its own which shows great resistance to being altered even by quite large external pressures. For example, a 2 mm diameter glass rod, of length 20 cm and supporting a weight of 1 kg, lengthens at a uniform rate of only 1 mm *per year* at a temperature of 500 °C, while at 300 °C the same elongation would take 100 years. Thus, while glasses behave to some extent like liquids, their viscosity is enormous even in comparison with very viscous liquids: the coefficients are some 10^{16} times larger than that of glycerine at 20 °C, and the viscosity increases greatly when the temperature falls.

An overall view of disordered structures: short-range order

So far in this chapter we have assembled a class of disordered structures which includes gases, liquids and *some* solids. Gases form a dispersed state of matter while liquids and amorphous solids are condensed states. We have classified them all together because it is always possible to pass continuously from one to another by a change of temperature or pressure. That, however, does not exclude the possibility of *also* observing a *discontinuous* transition between liquid and gas under certain limited conditions.

Let us clarify what we understand by disorder in a molecular arrangement. Disorder is complete if, around a given molecule taken as the origin of coordinates, the position of any other molecule is absolutely indeterminate. That is the situation for every molecule in a very dilute gas, i.e. a perfect gas. For substances with greater densities (compressed gases and, above all, liquids and amorphous solids), *disorder is complete only at larger distances* greater than a few molecular diameters (> 5). We say that there is *long-range* disorder. For pairs of molecules separated by shorter distances, the mutual position is not totally indeterminate because of geometrical conditions imposed by the interaction of neighbours or even their contact. However, these conditions are not enough to fix the arrangement with absolute rigidity, particularly as the distance apart increases beyond five or so molecular diameters. We say that, in these cases, there is *short-range order* which becomes more and more blurred as we move away from a given molecule and which disappears altogether at large distances.

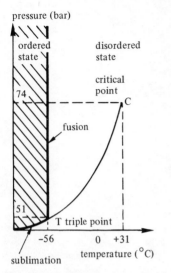

pressure (bar)

ordered state

disordered state

critical point

74

fusion

C

51

T triple point

−56   0   +31

temperature (°C)

sublimation

Fig. 2.6 Equilibrium phase diagram typical of all substances, but with values specifically for CO_2. The two lines labelled 'fusion' and 'sublimation' which meet at T separate ordered states from disordered ones. The vaporization curve is completely within the region of disordered states: it starts at the triple point T, where solid, liquid and gas coexist, and ends at the critical point C.

The ordered state: crystalline solids

The transition disorder ⇋ order

As we have seen, between certain solids and their corresponding liquids there is a continuous transition. For the great majority of substances, however, such a transition is discontinuous. In these cases, the change from solid to liquid is called **melting** or **fusion** and its converse is **solidification** or **freezing**.

For pure substances at a given pressure, these transitions occur at a definite temperature called the melting point or freezing point. For example, H_2O is a solid (ice) to within 0.001 K below 0 °C at atmospheric pressure, and is a liquid (water) to within 0.001 K above 0 °C. This discontinuous change of state is always accompanied by a change in specific volume (or density) and by the release or absorption of heat. There is also a discontinuous transition from gas to solid known as **condensation**, or from solid to gas which is called **sublimation**.

The equilibrium phase diagram of Fig. 2.6 gives the temperature of all these changes as a function of pressure for a typical substance, the values in the figure being specifically those for CO_2. The total area of the p–T plane is divided into two regions separated by the equilibrium line consisting of the fusion and sublimation curves which meet at T. This is a line of fundamental discontinuity because *it separates the ordered from the disordered states. It is absolutely impossible to pass from one region to the other without crossing this line*: it signifies an essential structural change. The region of ordered structures is that of **crystalline solids** or **true solids** as opposed to amorphous solids which we recall as being really only liquids with their disorder frozen in.

In a crystalline solid, the neighbours of every molecule are arranged in a regular pattern that is constant throughout the crystal. There is thus order over large distances, called *long-range order*. Even when two molecules are separated by hundreds or thousands of intermediate ones, their distance apart is governed rigorously by the demands of the particular crystal pattern. Crystals of silicon have been made with perfect order over as much as 1 cm and that is over 20 million atomic spacings.

We have already seen that by compressing a dispersed collection of molecules we can pass *continuously* to a condensed state of matter, but one which remains disordered over large distances. However, the transition from disorder to order can only be made by a *discontinuous* process. A picturesque and instructive example is provided by examining how a disordered heap of cobble-stones is transformed into a regularly paved surface (Fig. 2.7). It is no good pushing the heap this way and that and expecting it to become an orderly array. Instead, it is essential to start by arranging a small group of stones by hand according to the chosen pattern, and then to use this group as a 'seed', making it grow by adding carefully placed stones around it following the same pattern[1].

[1] The discontinuity between disorder and order does not exist for a one-dimensional arrangement. If identical balls are dispersed at random along a

Fig. 2.7 The transformation of a heap of cobble-stones from disorder to regularity. The 'discontinuity' at the time of the transformation is provided by the deliberate intervention of the paver, who arranges the stones one at a time according to the pattern to be achieved. (Craven-Rapho.)

Thermal motion in crystals

Once again, we have to say that the static model is not strictly correct because we must add the fact that molecules are always in thermal motion. The question which then arises is: how can we reconcile such irregular movements with the order that we have said exists in the crystal? To discuss this, we take the particular case of monatomic molecules where the argument can be made simple but, nevertheless, correct. Each atom is vibrating about a fixed point with an amplitude that increases with rise in temperature. *It is the average atomic positions which are perfectly ordered over large distances.* Thus, in solid neon, the distance between two atoms which are *nearest* neighbours will be:

$$a \pm \text{a fluctuating distance of the order of } 0.1a$$

at a temperature a little below the melting point. However, the distance between the first and the *millionth* atom will be:

$$10^6 a \pm \text{a fluctuating distance of the order of } 0.1a$$

and the variation here is negligible in comparison with the distance itself (Fig. 2.8). In other words, the centres of mass of the vibrating atoms are strictly ordered, and their vibrations are statistically the same.

The cohesive forces in the solid maintain the strict long-range order *in spite of atomic vibrations.* For every solid, however, there is a temperature at which the vibrations become so great that the

---

straight line so that their distances apart vary haphazardly, we have a 'one-dimensional gas'. If we gradually compress this array, the result is a collection of touching balls that are uniformly spaced and we have produced a 'one-dimensional solid' without discontinuity. This is impossible in three dimensions, so what is true in the one-dimensional case is not true in the three-dimensional case. This shows how dangerous it can be to reason from over-simplified models.

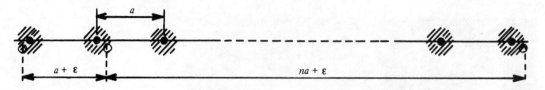

Fig. 2.8 Disorder due to thermal vibrations. The actual distance between two atoms differs from the theoretical distance (a, $2a$, $3a$, . . ., na) by an amount ε due to the vibrations. However, ε is the same for any two atoms, no matter how far apart, and therefore becomes negligible in comparison with na when n becomes larger.

structure suddenly breaks up: the solid melts. Why does this solid–liquid transition occur so abruptly over a temperature range that is practically zero, and how could we predict this temperature from the known properties of the atoms? These are questions we cannot answer in simple and accurate terms and they provide yet more examples of the limitations of structural models, at least when we are concerned with quantitative features.

Structure and properties

The various models that we have described in outline and have classified will be examined in more detail in the following chapters and we shall then show how *some* physical properties can be deduced from them. There are two simple cases from the theoretical point of view: that of complete disorder and that of perfect order. In the latter case, a small number of parameters is enough to describe the complete structure of the solid. In the case of total disorder, simplicity arises from the fact that macroscopic properties depend purely on average values and not on those for individual molecules. With complete disorder, a statistical calculation of average values can be carried out. As soon as even a partial degree of order occurs, the calculations become more complicated and even impossible. That is why the theory of liquids is not in a very advanced state, at least in comparison with that of perfect gases.

In fact, we begin with the case of perfect gases in Chapter 3 and then turn to the other extreme, that of crystalline solids, in Chapter 4. Later we shall have to accept that there is comparatively little that we can say at an elementary level about liquids and amorphous solids.

Between order and disorder

In the interests of clarity, we have insisted on a fundamental distinction between order and disorder. Such a sharp division is a very good first approximation and one that, without modification, enables us to explain many phenomena. However, there are other phenomena, no less important, whose explanation does need some modification of our first idea and will unfortunately make it more complicated.

Thus, in the same way that a little short-range order was introduced into the disordered structure of liquids, so *real* crystalline solids will be found to contain some disorder. This occurs in the form of irregularities and imperfections in the basic structure. While they may be quite rare or quite modest, they can have a profound influence on physical properties so that a considerable proportion of the chapters devoted to crystalline solids will deal with models of imperfections in perfect crystals.

In addition, there exist some materials with a structure that is genuinely intermediate between order and disorder. They are clearly special cases, but they must not be ignored just because they are difficult to classify. In fact they have interesting properties which result directly from the complexity of their structures and our knowledge of them has recently made great progress. They can be classified as poorly crystalline or as partially crystalline, and include long-chain polymers (plastics) and many substances that are essential to our own bodies. The category also includes **liquid crystals**, whose very name seems to challenge our 'fundamental' distinction between crystal and liquid. We shall, however, examine all these in Chapter 9 only after deepening our understanding of the simple and well-defined concepts of order and disorder.

The universality of structural models

As a general rule, *all* substances can exist in three states or phases (i.e. crystalline solid, liquid or gas) which are separated by equilibrium lines of the same shape as those in the phase diagram of Fig. 2.6. The only differences that occur from one substance to another are the p and T coordinates of the triple point T and the critical point C. Whether we usually think of a substance as a solid, a liquid or a gas depends only on the position in the phase diagram of the point corresponding to ordinary temperatures and pressures—that is, about $T = 293$ K and $p = 10^5$ Pa or 1 bar.

When we consider the great diversity exhibited by matter in the actual world around us, it was a remarkable scientific advance to achieve such a universal view of the behaviour of substances. The decisive step in this advance was the liquefaction of certain gases which had always been considered as 'permanent', a process that needed methods of lowering temperatures nearer and nearer to the absolute zero. The first liquefaction of a 'permanent' gas, that of oxygen, was only carried out about 100 years ago in 1877 by both Cailletet and Pictet.

We shall restrict the range of temperatures and pressures over which we study the states of matter to those conditions normally attainable in the laboratory—from just above 0 K to about 4000 K for temperature and from just above zero to a few GPa (a few tens of kilobars) for pressure. This means that we do not intend to look at matter in states that may exist at the centre of the earth or in stars, where temperature and pressure can be considerably higher. Under the restricted conditions that we shall be exploring, it is possible that an occasional phase may not exist: thus, a liquid form of carbon is not known.

Finally, a few special states lie outside the scope of this book. Firstly, we shall not study *plasmas*. These are non-equilibrium gaseous states in which the 'particles' are charged ions and electrons created by intense electric fields producing discharges in a gas. Secondly, we shall not study *helium* at very low temperatures. Above a certain critical temperature (5.21 K for the ^{4}He isotope,

3.38 K for the ^{3}He isotope), helium behaves like a normal gas; below this temperature, where it is in a condensed state, its properties are different from those of every other substance—there is no triple point and a superfluid phase of the liquid exists. These peculiarities stem from the fact that the interactions between all other atoms are governed largely by the laws of classical physics to a high degree of accuracy, whereas the behaviour of helium atoms deviates widely from what is predicted by classical laws and is governed by *quantum statistics*.

3 The perfect gas

The model we have arrived at for the structure of a perfect gas consists of molecules *in complete disorder moving with quite random speeds and directions.* In addition, we know that *the energy of this molecular movement increases with rise in temperature.* We shall see in this chapter how, from these simple qualitative statements, the kinetic theory of gases enables all the laws to be deduced which accurately describe the behaviour of these gases—laws which are completely verified by experiment.

The kinetic energy of molecules

The mean energy and the temperature scale

We must first determine exactly how much we can find out about the magnitudes of molecular speeds. Let us imagine that we are following the path of a given molecule: it travels along a straight line at speed v_1 until it encounters another molecule; the collision changes its direction and gives it a speed v_2; then another collision occurs, giving it a speed v_3, and so on. We shall see later that, for a gas under normal conditions, the molecule undergoes something like several thousand million collisions per second, so that the changes in velocity are extremely rapid. Along each of the straight segments of the path, the kinetic energy of the molecule is $\frac{1}{2}mv_1^2$, $\frac{1}{2}mv_2^2$, $\frac{1}{2}mv_3^2$, etc. It clearly takes on a large number of successive values which fluctuate around a mean value E_m that can be written

$$E_m = \tfrac{1}{2}mv_m^2$$

The quantity v_m is the speed of the molecule whose kinetic energy would be E_m and it is called the mean molecular speed. If, instead of following a particular molecule over a period of time, we recorded the speeds of all the molecules in a vessel at the same instant, we should find the same value for the mean kinetic energy E_m.

This mean kinetic energy is characteristic only of the temperature of the gas: for a given temperature it has a fixed value which *does not depend on the nature of the gas*—a remarkable fact that we shall return to later.

Thermal motion, or more exactly the quantity $\frac{1}{2}mv_m^2$, always increases as the temperature rises, so this quantity can be used without ambiguity to define a temperature scale. It is simplest to make E_m proportional to the temperature T by putting

$$E_m = \tfrac{1}{2}mv_m^2 = AT$$

where A is an arbitrary constant. By fixing the value of A the numerical value of the temperature T is also fixed and thus defines what is called a 'scale of temperature'.

The present universally adopted convention is to choose the coefficient A in such a way that, at the triple point of water[1], T is equal to 273.16. The scale is then the *absolute thermodynamic scale* whose unit is the **kelvin** K. The choice of this rather strange number appears arbitrary, but in fact there is a good reason for it: in the temperature scale defined in this way, the melting point of ice at atmospheric pressure is 273.15 K and the boiling point of water at the same pressure is 373.15 K, exactly 100 degrees greater. The ordinary everyday temperature scale (the celsius scale) can thus be obtained from the absolute temperature by subtracting 273.15 from it:

$$t \, (°C) = T \, (K) - 273.15$$

although we often use 273 as a rounded-off figure. In addition, as a *temperature interval*, the degree celsius is equal to the kelvin.

What value does this give to the coefficient A which defines the kelvin scale? First of all, it is generally written in the form $A = 3k/2$, where k is *Boltzmann's constant*: in other words, the absolute temperature scale is defined by the equation

$$\tfrac{1}{2}mv_m^2 = 3kT/2$$

Boltzmann's constant will have a unit corresponding to that of energy per unit temperature increase. In SI units, it is found that

$$k = 1.38 \times 10^{-23} \, \text{J K}^{-1}$$

or, using the electron-volt as the unit of energy,

$$k = 0.86 \times 10^{-4} \, \text{eV K}^{-1}$$

It is useful to remember the following convenient values for the quantity kT, which also help in recalling the value of k itself: at ordinary temperatures (about 300 K), kT is 1/40 eV, and at 1200 K it is 0.1 eV.

The absolute temperature is thus a measure of the energy of the thermal motion of molecules. Quite clearly, it can only be positive, so that the *zero of the kelvin scale* (the absolute zero) is the limit of low temperatures[2].

The distribution of molecular speeds

Not only is the *mean value* of the kinetic energy per molecule known for a gas at temperature T, but the *distribution* of energy among the individual molecules is known as well. The theory of

[1] There is only one temperature and pressure at which ice, water and water vapour can coexist in equilibrium: the triple point T in Fig. 2.6.
[2] This method of defining the temperature scale introduces the idea of a lower limit to temperature in a very natural way. This would not be the case if we had started from the temperature scale based on the fixed points of water. In that case, we should have had to assert that it is impossible to lower the temperature below −273.15 °C, something that would appear surprising to a newcomer to the subject.

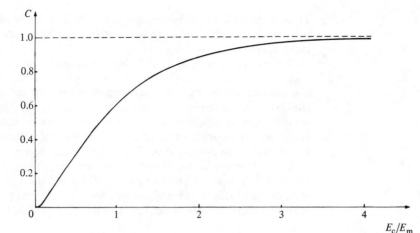

statistical mechanics enables us to calculate what fraction of the total number of molecules have energies in any particular range of values.

Figure 3.1 gives the results of these calculations and shows that there is quite a spread of energy values. The following figures, valid for all gases at all temperatures, can be obtained from the curve: 32% of the molecules have an energy less than half the mean E_m, 57% have energies between $\frac{1}{2}E_m$ and $2E_m$, while the remaining 11% have energies greater than $2E_m$. In this last group, less than 1% have energies in excess of $4E_m$.

The figure also shows that 52% of the molecules have speeds that differ from the mean speed by less than 30%. For, in the range of speeds from $0.7v_m$ to $1.3v_m$ (i.e. a range of E_c from $0.49E_m$ to $1.69E_m$), there are 52% of the total, as the reader can check from the curve.

Thus, whereas the positions of the molecules and their directions of motion are completely random, the absolute values of their speeds are not, but tend to cluster around the mean value. The preceding paragraphs show that it is quite a crude approximation to take the mean value of $3kT/2$ as the energy at any instant of a single molecule, but the mean value itself is an essential factor in an exact specification of the model for a perfect gas. This is something we have indicated before, but it is extremely important and we emphasize it again.

Whatever the nature of the molecule, its mean kinetic energy is the same at any given temperature. Thus our model of a perfect gas, consisting of a collection of disordered, moving 'material points', is universal provided that the parameters used are the *position* of the point, the *direction* of its motion and its *kinetic energy* E_c. Such a model is applicable to all gases in the perfect state or in a state very close to it.

Different perfect gases are distinguished from each other only by the mass m of the material point representing the molecule. It follows that the speeds of thermal motion in two different gases are not the same because

$$v = \sqrt{(2E_c/m)}$$

Fig. 3.1　Theoretical curve giving the proportion C of gas molecules with a kinetic energy less than a given value E_c. The energy E_c is expressed as a fraction or multiple of the mean energy per molecule E_m.

Thus, even with the same energy, a molecule which is four times the mass of another has only half the speed. This is also true for mean values.

This perfect gas model, and the consequences we shall draw from it, are only strictly valid for a real gas when the density is low enough for interactions between molecules to be negligible. However, this criterion is too vague to give us any general rule for the maximum number of molecules per unit volume permitted if the model is still to apply. The values of density and temperature at which deviations from the perfect state become appreciable vary with the structure of the molecule and thus with the type of gas. Thus, even at the same temperature (600 K) and the same volume available per molecule (1 nm³), argon deviates from the perfect gas laws by about 1% while water vapour deviates by 20%.

The perfect gas laws

Having established the model, we now ask what conclusions can be drawn from it. What predictions will it make about the macroscopic behaviour of a perfect gas that can be checked experimentally? First of all, we shall show how the pressure of the gas is related to the mean molecular kinetic energy and this will yield the well-known 'perfect gas law' or equation of state.

The equation of state for a perfect gas: $PV = nRT$

Consider n moles of a gas (containing Nn molecules, where N is Avogadro's number which is defined on p. iv of the Preface) confined to a volume V at a temperature T by a piston as in Fig. 3.2. From the macroscopic point of view, this confinement requires the application of a pressure P to the piston which exactly balances the pressure of the gas on the lower face. *The kinetic theory of gases* enables us not only to *understand* how this latter pressure arises but also how to *calculate* it accurately.

The fundamental cause of the pressure is the collision between molecules and the surface of the piston. Consider first a rather unreal situation in which all the molecules have velocities normal to the face of the piston and the same speed of impact v. We know already (p. 15) that atoms are impenetrable and behave like hard spheres, so the molecules will rebound from the piston. The collisions, moreover, are *elastic*, which means that the molecules and the piston after collision are left intact and the molecules lose no kinetic energy. Their speed is thus the same after impact as before, although of course it is in the opposite direction. Although the kinetic energy is constant, the *momentum* changes from $+mv$ to $-mv$, a total change of $-2mv$. According to Newton's second law of motion, this change must be produced by a force exerted *by* the piston *on* each molecule—and by Newton's third law, the same force is exerted *on* the piston *by* each molecule.

The number of molecules hitting the piston per second is extremely large: for example, in nitrogen under normal conditions it is of the order of 10^{24} per second per cm². Given such a high

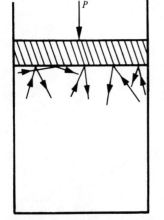

Fig. 3.2 Collisions of molecules with the walls of the vessel are the cause of the pressure exerted by the gas.

frequency of collision, the total effect of all the collisions is apparently constant and this effect produces the upward pressure on the piston in Fig. 3.2 that we need to calculate.

The total pressure required is the *force* per unit area, and the force is given by the change in momentum per second, by Newton's second law of motion. Since the change in momentum in total is given by the product (*number* of collisions × change of momentum in *each* collision), we can write:

pressure P = force per unit area
　　　　= change of momentum per second per unit area
　　　　= (number of collisions per unit area per second)
　　　　× (change of momentum in each collision)

The first factor here is proportional both to the speed v and to the density of the gas or the number of molecules per unit volume Nn/V. The second factor, as we have already seen, is $2mv$. Collecting together these expressions, we see that the pressure is given by

$$P \propto (nN/V)(mv^2)$$

Earlier in this chapter, we showed that in general the molecular kinetic energy $\frac{1}{2}mv^2$ is proportional to the absolute temperature T, so we conclude finally that

$$P \propto (nN/V)T$$

In fact, an exact calculation with this unreal model gives $P = (nN/V)\,3kT$.

If we now pass from the simplified model to the real situation, where the molecules have a distribution of speeds and move in all directions[1], an exact calculation shows that the results obtained above are still valid as far as the *dependence* of P is concerned: it is still proportional to $(nN/V)T$. In spite of the complications in dealing with the real case, it is also possible to determine the coefficient of proportionality exactly and it is found to be simply the Boltzmann constant k, instead of the $3k$ yielded by our unreal case.

Finally, therefore, the kinetic theory of gases establishes that a perfect gas obeys the following law, called its *equation of state*:

$$P = (nN/V)kT$$

or, in other words, either

$$PV = nNkT = nRT$$

or　　$PV_{\mathrm{m}} = RT$

[1] The velocities of the molecules are in fact distributed among all directions with equal probability: in other words, perfectly isotropically. In that case, it can be shown that their average effect is the same as if they were in fact moving in equal numbers in any three mutually perpendicular directions. If one of these directions is chosen to be along the normal to the piston, so that the two others are parallel to it, it is easy to see that the pressure would be the same as if only one-third of the total number of molecules struck it.

in which we recall that

(1) n is the number of moles contained in the volume V at temperature T under a pressure P.

(2) R is the *molar gas constant* equal to Nk, where N is Avogadro's number and k is Boltzmann's constant. Its value is

$$R = 8.31 \text{ J K}^{-1} \text{mole}^{-1}$$

(3) V_m is the volume of 1 mole, the *molar volume*.

The equation of state accounts completely for the behaviour of a perfect gas as described by the various laws. For instance:

Boyle's law: at constant temperature, PV for a fixed mass of gas is constant.

Expansion at constant pressure: if V is the volume at temperature T and V_0 the volume at 0 °C (T_0), then $V/V_0 = T/T_0$: that is, the volume is proportional to the absolute temperature if the pressure is constant. With the temperature expressed in degrees celsius (t), this law takes the form $V = V_0 (1 + t/273)$.

Charles's law: at a constant volume, the pressure of a gas is proportional to the absolute temperature or $P/P_0 = T/T_0$, where P_0 is the pressure at 0 °C. In terms of t, $P = P_0 (1 + t/273)$.

Avogadro's law: equal volumes of all gases at the same temperature and pressure contain the same number of molecules. This number, nN, is equal to PV/kT, showing clearly that the law is valid.

The equation of state is a direct consequence of the model adopted for the perfect gas. As long as the conditions specifying the perfect state are fulfilled (i.e. very low density, no interaction between molecules), the equation is found to hold exactly, but the more we depart from these conditions, the more it becomes an approximation. This is illustrated in Fig. 3.3. Of course, the equation fails completely when the gas transforms to another state of matter, whether liquid or solid. At a given pressure, it is only applicable above the vaporization temperature and even then only approximately at first, although the agreement becomes better and better as the temperature rises.

Heat capacity of a perfect gas

We now give another example of simple results that can be deduced from the model of a perfect gas. However, whereas the equation of state was valid for all gases, we must restrict ourselves in this example to those with monatomic molecules, that is, to inert gases or metallic vapours. Consider a mole of the gas contained in a vessel of *fixed volume* independent of the temperature[1]. There is no interaction between the molecules of a perfect gas, so that raising

[1] The variation in the volume of the solid vessel itself due to thermal expansion is negligible.

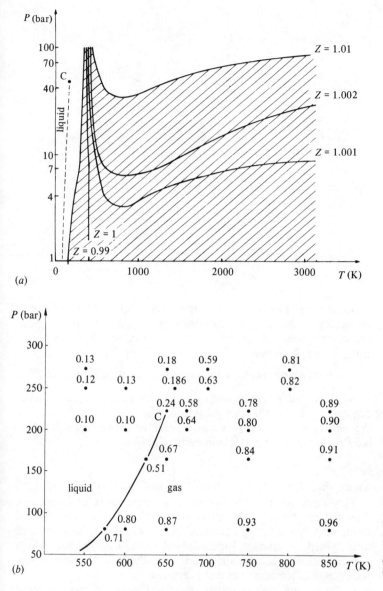

(a)

(b)

Fig. 3.3 (a) Values of the quantity $Z = PV_m/RT$ as a function of pressure P and temperature T for argon. If $Z = 1$, the gas is perfect. In the shaded region, Z differs from 1 by less than 1%. The gas departs considerably from the perfect state at high pressures and in the neighbourhood of the vaporization or condensation curve terminated by the critical point C. (b) Values of $Z = PV_m/RT$ for water as a function of P and T around the critical point C. Departures from $Z = 1$ become considerable near the condensation curve and above the critical pressure.

the temperature of the gas succeeds only in increasing the kinetic energy of the molecules and nothing else.

Let us examine this last point more carefully. Suppose the gas is contained in a vessel closed by a movable piston. When the gas is heated its volume increases, the piston is moved, and so the gas does a certain amount of work. A quantity of energy equivalent to this work must be given to the gas. However, if the vessel is rigid, the volume is fixed and so that effect does not occur.

There is, in addition, another aspect. We have mentioned several times that the molecule is to be regarded as a 'material point', but in this present calculation that assumption is only valid for molecules consisting of a single atom as we shall discuss later.

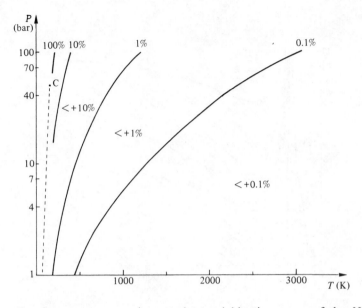

Fig. 3.4 Molar heat capacity at constant volume for argon. Values given are percentage deviations from the theoretical value $C_v =$ 12.46 J K^{-1} mole^{-1}. The point C is the critical point at the end of the condensation curve.

So, for our monatomic gas, the total kinetic energy of the N molecules in the mole of material is NE_m, since E_m is the mean energy *per molecule*. According to the relation on p. 38, $E_m = \frac{1}{2}mv_m^2 = 3kT/2$, so that

total kinetic energy of a mole of gas $= N(3kT/2)$ or $3RT/2$

Now the *molar heat capacity* of a substance at constant volume, denoted by C_v, is the *energy absorbed by a mole of it for every degree rise in temperature*. Clearly, from the above expression, this is $3R/2$ so that for a perfect monatomic gas

$$C_v = 3R/2$$

which, since $R = 8.31$ J K^{-1} mole^{-1}, gives $C_v = 12.46$ J K^{-1} mole^{-1}.

This value for C_v is *independent of temperature and pressure*. The theory, without the introduction of a single experimental parameter, has given a numerical result which can be compared directly with experiment. Figure 3.4 shows the accuracy with which the theoretical predictions are verified in the case of argon over a wide range of temperatures and pressures.

If the gas molecule consists of several atoms, things are more complicated. Thermal movement in a monatomic molecule can only mean translation: that is, movement of the whole molecule along a trajectory. That is the case we have just dealt with. However, if the molecule is polyatomic, the movements of the atoms can be analysed into a *translation* of the whole molecule, a *rotation* of the molecule about its centre of mass, and a deformation of the molecule by the *vibration* of its atoms around their mean positions (Fig. 3.5). So when the temperature is raised, the heat absorbed goes to increasing *not only* the kinetic energy of translation *but also* the energies of rotation and vibration. This means that the molar heat capacities of polyatomic gases can be *greater than those of monatomic gases* but in any case cannot be

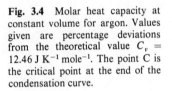

translation of the whole molecule

deformation of the molecule by vibration

rotation about an axis

Fig. 3.5 Analysis of the movements of a diatomic molecule.

Table 3.1 Values of molar heat capacities at constant volume for several monatomic and polyatomic gases.

Substance	Molecule	Molar heat capacity at constant volume ($J\,K^{-1}\,mole^{-1}$)	
Argon	Ar	12.46	Monatomic
Oxygen	O_2	20.8	Diatomic
Nitrogen	N_2	20.6	(note 12.46 ×
Carbon monoxide	CO	20.6	$5/3 = 20.7$)
Hydrogen	H_2	20.1	
Water vapour (100 °C)	H_2O	25·4	
Carbon dioxide	CO_2	28.0	
Nitrogen dioxide	NO_2	28.1	Polyatomic
Ethane	C_2H_6	39.6	
Hydrogen sulphide	H_2S	27.4	

smaller. This is confirmed by experiment as the values in Table 3.1 show.

It is found that several diatomic gases have the same heat capacity, one that is in the simple ratio of 5/3 to that of the monatomic gas. Moreover, the heat capacity of molecular hydrogen H_2 varies with temperature, and below 40 K reduces to the same value as that of a monatomic gas. These results are not just chance, but can be explained quite simply using the ideas of classical and quantum statistical mechanics, although we do not wish to introduce these into the present book[1].

For very complicated molecules, it is no longer possible to calculate the heat capacities from a knowledge of their structure. This is much the same situation as with a gas that is no longer perfect because its molecules are not strictly independent of each other: the complexity makes exact calculation impossible.

The examples given in the last two sections are typical of those that can be explained using molecular models. The kinetic theory of gases and statistical mechanics together allow us to predict various easily measured macroscopic properties exactly *without the introduction of experimental data* but from first principles. In this way we have calculated the pressure of a given mass of gas at a given temperature occupying a given volume, the heat energy needed to raise the temperature by so many degrees, and so on.

The essential fact to keep in mind, however, is that the calculations are made possible and simple *only because the gas is perfect* or, what amounts to the same thing, *disorder is complete*. With a real gas whose density is not low enough, rigorous

---

[1] For the benefit of readers who are aware of these theories, we recall that:

(1) When the theorem of the equipartition of energy is valid, translation involves three degrees of freedom and rotation involves two, so the energy of thermal motion is $3RT/2$ for monatomic gases and $(3 + 2)RT/2$ for diatomic.

(2) For O_2, CO_2, etc., the energy of vibration is not involved because its quantum is too large compared with kT at ordinary temperatures. For H_2, the quantum of energy of rotation is also too large below 40 K, so the rotation of the molecule is not excited and it behaves like a monatomic gas.

calculation is no longer possible because the laws of interaction between molecules are not known with sufficient accuracy. Moreover, the calculations become very complicated because, unlike the case of the perfect gas, the internal structure of the molecule is involved.

Collisions between molecules

Until now, it has been assumed that collisions between molecules were merely the means by which they became a disordered equilibrium collection. Neither the mechanism of the collisions themselves nor their frequency has come into question. We are now going to examine these aspects more carefully and this will produce further links between molecular behaviour and measurable properties.

The mean free path

A molecule follows a complicated path consisting of sections between collisions that are straight lines, each section being joined to the adjacent ones at random angles. The whole trajectory resembles a broken line rather like that illustrated in Fig. 2.4 for Brownian motion. The length of each section is determined by the positions of two successive collisions and is called the *free path* between the collisions. The lengths of the sections are variable but they fluctuate about a mean value l called the **mean free path.**

For the purposes of our simple arguments leading to the perfect gas laws, we assumed that all the molecules had a speed equal to the mean value v_m defined on p. 37. We showed in fact that although this approximation was quite crude it was not completely absurd. In any case, it does allow us to evaluate the mean time between collisions given by $\tau = l/v_m$. The mean frequency of collisions is then $1/\tau$.

What factors determine the size of these quantities? It is immediately clear that a model with molecules represented by 'material points' is certainly inadequate and cannot tell us anything since its geometry is non-existent. So once again we resort to the next simplest case, that of an inert gas in which a molecule is represented by a hard sphere of diameter d. Suppose a molecule A (Fig. 3.6) has just experienced a collision at O and is now moving along Ox. A new collision will occur when A encounters another

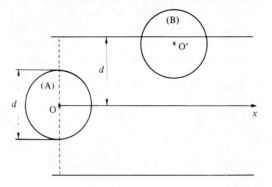

Fig. 3.6 Collision between two molecules. Any molecule whose centre O′ is inside the cylindrical tube of radius d will suffer a collision from A. The area πd^2 is called the collision cross-section.

molecule *whose centre is inside a cylindrical tube of radius d and axis Ox*. So if the distance from the centre O' of a second molecule B to Ox is greater than d, A and B will not collide. If O' is slightly less than d from Ox, a glancing collision will occur and A will be slightly deviated from its trajectory along Ox. If, at the other extreme, O' lies on Ox itself, the collision would be head-on and A would recoil in the opposite direction along xO.

If, on average, the next collision suffered by A takes place when its centre is at a point we call O_1, say, the distance OO_1 is the *mean free path*. It is easy to see that OO_1 will become shorter as the density of molecules increases. In addition, it will also get shorter as the cross-section of the cylinder within which the centre of B must lie for a collision to occur becomes larger—more simply, larger molecules have shorter free paths, as any plump person struggling through a crowd can testify. These arguments lead to the conjecture that the mean free path l is inversely proportional to n, the number of molecules per unit volume, and to the cross-section $\sigma = \pi d^2$. In fact, if an exact calculation is made on the assumption that all the molecules have the same speed, the formula yielded for l is:

$$l = 1/\sqrt{2}n\sigma$$

Molecular dimensions are known for simple cases (see Table 1.5), so it is possible to determine an order of magnitude for the mean free path. Thus, for argon under normal conditions,

$$n = 2.7 \times 10^{25} \text{ molecules per m}^3$$
$$\sigma = \pi \times (0.384 \times 10^{-9})^2 \text{ m}^2$$

and these give

$$l = 0.06 \text{ μm}$$

From the relation $\frac{1}{2}mv_m^2 = 3kT/2$ (p. 38), the root mean square speed of a molecule is given by

$$v_m = \sqrt{(3kT/m)}$$

and, for argon at 300 K, has the value 430 m s^{-1}. Thus the mean time between collisions $\tau = l/v_m$ is 0.15×10^{-9} s or 0.15 ns. The molecule suffers around eight thousand million collisions per second and can only travel on average 65 nm (180 times its own diameter) before colliding with another molecule and changing its direction. These figures give some idea of the chaotic disorder at the molecular level, at the same time as the gas appears perfectly calm on a large scale.

To reduce the frequency of collision, the density n can be lowered by a reduction of pressure—this is often called 'creating a vacuum'. For a pressure of 10 Pa, l is of the order of a mm; at 0.1 Pa it is more than 10 cm. Pressures of 0.1 Pa are easy to obtain with a good vacuum pump. (The word 'vacuum' is a relative term, since at this pressure there are still 2.5×10^{19} molecules per cubic metre.)

Thus when a vessel of ordinary size (about 10 cm) contains a gas at a pressure below 0.1 Pa, the molecules on average do not

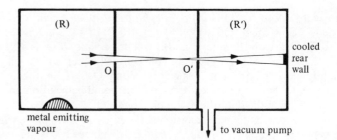

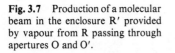

Fig. 3.7 Production of a molecular beam in the enclosure R′ provided by vapour from R passing through apertures O and O′.

collide with each other at all but only with the walls of the vessel. Under these conditions, the disorder is very different from that in a normal gas, as are the physical properties. This state is called the *molecular regime* and the flow of a gas at these pressures is called *molecular flow*, or *free-molecule flow*.

Molecular beams

As an important example of these conditions, we look briefly at the case of *molecular beams*. Suppose we take a vessel containing a gas at low pressure, produced for instance by a metal vapour in equilibrium with its solid form at a temperature which can be adjusted according to the pressure required. This vessel (R in Fig. 3.7) communicates with a second enclosure R′, evacuated as completely as possible, through two small apertures O and O′. The molecules which escape from R have a mean free path greater than the length of R′ and are so directed by the apertures O and O′ that they reach the back wall of R′ without any collisions. The wall surface is cooled to such an extent that the molecules hitting it condense on it and stick to it without being reflected. The enclosure R′ is thus traversed by a jet of gas consisting of a fine beam of molecules with well-defined trajectories. The speeds of the molecules which leave R have a distribution corresponding to the temperature in R, but the selection of molecules by the apertures O and O′ eliminates the disorder in *direction* which is characteristic of normal thermal motion.

Beams like this are used in various experiments. For example, the *speeds* of the molecules can be measured as they move along straight paths in R′ and this can be used as an experimental check of the theoretical law of distribution given in Fig. 3.1. For this, the molecular beam is interrupted by a disc rotating at a great speed but with a slit in it (Fig. 3.8) allowing short pulses of molecules to pass through it. At a distance L away, a second identical disc is fixed on the same axis of rotation and has the same rotational speed as the first, but the slit in it is displaced through an angle α as shown. A pulse of molecules will pass through both slits only if its speed carries it over the distance L in the time t_0 taken for the discs to turn through α. The time t_0 is given by the value of α divided by the angular speed of the discs, and the molecular speed is then L/t_0. Only molecules with this speed will get through both apertures. (This is the same principle as that of synchronized traffic lights which allow continuously free passage to cars moving with a

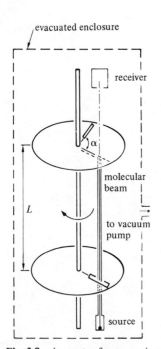

Fig. 3.8 Apparatus for measuring molecular speeds in a molecular beam.

certain speed.) The apparatus thus classifies molecules according to their speeds. The relative number of molecules in each small range of speeds can be assessed by finding the time taken for the deposit on the back wall to become just visible, although there are more sophisticated methods. The distribution of molecular speeds can thus be ascertained.

Macroscopic phenomena dependent on mean free path

On p. 47 we evaluated the mean free path l in terms of molecular dimensions by adopting a simplified model of collisions between hard spheres of given diameter. The formula can also be used, conversely, to find the value of σ, the *effective* collision cross-section, provided that the parameter l can be experimentally determined. The quantity σ is characteristic of the geometry of the molecule and of the collision conditions, but its definition is not quite straightforward: the length d in $\sigma = \pi d^2$ is the diameter of hard spherical molecules which at the same density would have the same free path as the given molecules, whatever their shape. Our interest in the collision cross-section arises from the fact that there are measurable properties that *will* give us the mean free path experimentally. Two of these are viscosity and thermal conductivity.

The viscosity of gases

We have already introduced, even if only qualitatively, the idea of viscosity in liquids. On a much feebler scale, a gas, even one that is perfect, has a certain viscosity which becomes evident when layers of it slide over each other at different speeds. The simplest model is illustrated in Fig. 3.9. The gas is contained between two parallel plates, one of which, P_1, is fixed and the other, P_2, is moving with a speed V in the plane of the plates. As a result of molecular collisions on P_2, the layer of gas in contact with it is dragged along. So the layer of gas in contact with P_1 is at rest and that in contact with P_2 is moving with a speed V. From P_1 to P_2 successive layers have speeds that grow linearly from zero to V.

If the layers of gas sliding over each other did not interact, there would be no resistance to the motion of P_2. In fact, the displacement of P_2 *does* need the exercise of a force proportional to its surface area. At the same time, an equal and opposite force is also needed to keep P_1 fixed in place. In principle, it is the measurement of these forces that yields the value of a coefficient giving a precise indication of the size of these effects. It is called the *coefficient of viscosity*, η. The viscosity of air is the main cause of the resistance to the movement of aircraft through it and also of the friction which produces the considerable heating of Concorde's fuselage.

The kinetic theory of gases enables us to evaluate this coefficient from calculations at the molecular level. On a molecular scale what

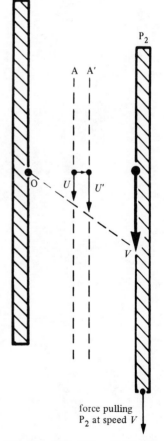

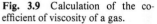

Fig. 3.9 Calculation of the coefficient of viscosity of a gas.

happens is this: a molecule in equilibrium in a certain layer of air A moving with a speed U is transported by its thermal motion to a neighbouring layer A' with a different mean speed U' (Fig. 3.9). The mean speed of the molecule is changed from U to U' by molecular collisions in the second layer and this implies a transfer of momentum from the layer to the molecule. To do this, the layer has to provide impulses to the molecule, thereby causing the layer to experience a dragging force. Molecules moving from A' to A will similarly transfer momentum so as to pull the slower layer along. The sum of all these transfers of momentum is the origin of the drag that each layer experiences from its slower neighbour and of the pull exerted by its faster neighbour. These drags and pulls produce the viscosity of the gas.

With a few simplifying assumptions, the calculation yields the following formula for the coefficient of viscosity:

$$\eta = 0.53 \sqrt{(mkT)}/\sigma$$

We quote this because some interesting consequences follow from it. Firstly, we take the case of argon which fits the hard sphere molecular model most closely. The effective molecular diameter d, calculated by obtaining σ from viscosity measurements and then using $\sigma = \pi d^2$, is 0.364 nm. This is in excellent agreement with the atomic diameter obtained by crystallographic methods and listed in Table 1.5 as 0.384 nm. For nitrogen, viscosity gives a diameter of 0.375 nm, which should be compared with the dimensions given in Fig. 1.10. Finally, we quote the effective diameters for O_2: 0.375 nm; for CO_2: 0.459 nm; for CH_4: 0.414 nm. It is remarkable that *the kinetic theory of gases gives values for molecular sizes in good agreement with those yielded by models with completely different bases.*

Notice that the pressure of the gas does not appear in the formula for viscosity, but only the temperature. This is a surprising result since we would expect frictional effects to be greater at a given temperature as the density of the molecules increases. Nevertheless, it is a result that is confirmed by experiment. Once we know that this is so, we can find an explanation: when the pressure is lowered, the number of molecules transported from one layer to another decreases, but the effect of each one is greater because the mean free path is longer. The interaction thus makes itself felt between layers that are further apart and therefore with greater differences in speed. The two effects compensate each other—at least until the mean free path becomes comparable with dimensions of the apparatus in the experiment, for example with the distance between P_1 and P_2 in Fig. 3.9. An abrupt change of properties then occurs because we are in the region of molecular flow mentioned on p. 48.

Another consequence of the formula is that the viscosity of a gas increases with rise in temperature, whereas common experience tells us that the viscosity of liquids decreases: they flow more easily. However, the basic causes of viscosity in the two disordered systems are quite different. In liquids, the molecules are in contact and the viscosity depends on the frictional forces between them which become smaller with the growth of thermal motion. In gases, the molecules are free and the interaction is at a distance, achieved through instantaneous collisions whose effect increases as the temperature rises.

Thermal conductivity of gases

Gases are good thermal insulators. Nevertheless, if a hot plate is separated from a parallel cold one by a layer of gas, there is a flow of heat from the hot surface to the cold one. This occurs even if all mass movement of the gas (convection) between the plates is prevented, as it must be when we want to examine conduction since convection is often the most important process by which heat is lost from a surface. To prevent convection in an experiment with hot and cold plates as in Fig. 3.10, the hot plate should strictly be horizontally above the cold one since otherwise the heated gas will rise and the cooler gas will fall. If, however, convection is avoided, it is found that the flow of heat by conduction through the gas between the plates is proportional to the temperature gradient: that is, the temperature difference divided by the distance apart of the plates. The constant of proportionality is called the *coefficient of thermal conductivity, K*, and it is a property characteristic of the gas. K can be measured directly by experiment but, once again, the kinetic theory of gases will give an expression for it based on a molecular model.

The successive layers of gas between the plates in Fig. 3.10 have temperatures that increase regularly as P_2 is approached. Thus the molecular motion in layer A' is greater than that in A. When a molecule jumps from one layer to a warmer one, from A to A' say, it finds itself surrounded by molecules which are, on average, moving more rapidly. As a result of collisions, the difference quickly disappears but the 'hot' molecules have given some of their energy to the 'cold' one—this is a transfer of heat, although it is in fact a transfer of kinetic energy, which is what the theory calculates. It can be seen that this mechanism has a certain analogy with the transfer of momentum which occurs in the case of viscosity. Indeed, theoretically there is found to be a very simple relation between the coefficients of viscosity (η) and of thermal conductivity (K): a simplified calculation shows that the ratio η/K is equal to the heat capacity of the gas C_v. Experimental results give a ratio that is rather nearer $2C_v$ but, given all the approximations used in performing the calculations, the agreement is not at all bad.

Like the coefficient of viscosity, that of thermal conductivity is independent of pressure until the molecular regime is reached at a high vacuum. A dewar (or 'thermos') flask (see Fig. 7.1) is a container whose contents are insulated by having two outer walls separated by a few millimetres. The insulation is not improved at all by lowering the air pressure in the gap from atmospheric (about 10^5 Pa) to 100 Pa, but at 1 Pa, which brings the air into the molecular regime, loss of heat by conduction through the air is eliminated.

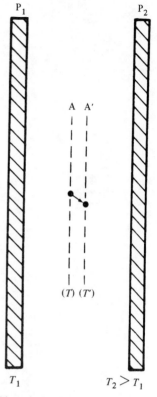

Fig. 3.10 Calculation of the thermal conductivity of a gas.

It is true that in this chapter we have concentrated on certain particular properties of perfect gases, but this is mainly because we wished to show how valuable a good model can be. Using quite elementary methods, we have discovered quantitative relationships between apparently very different phenomena. The experimental verification of these relationships, even if sometimes approximate, is fundamental because *it cannot be the result of chance.*

Moreover, these examples have also shown how macroscopic measurements can yield information about the molecular model used. When several completely independent measurements lead to closely similar values for molecular quantities (for instance, diameters) confidence is generated in the validity of the model used. It is true that details of the theories may be modified, sometimes even quite substantially, and that some relatively crude calculations may be refined, but beneath all these changes something always remains of the basic essentials of the original models.

Fluctuations in the density of gases: the scattering of light

On average, the distribution of $\mathcal{N}$ molecules over the volume V occupied by a gas is uniform owing to the complete disorder in position. The *mean* number of molecules in any small region of volume ΔV is given by $N_0 = (\mathcal{N} \, \Delta V)/V$. However, precisely because the disorder is maintained by thermal motion, the number of molecules in ΔV is not *exactly* N_0 at *every instant*. In fact, the actual number N in ΔV fluctuates about the mean N_0 so that

$$N = N_0 + n$$

where n is the *deviation* or difference from the average. Instead of considering the number of molecules in a fixed volume as time goes on, we could equally well take the number of molecules *at a definite instant* in *different* small volumes of the same size ΔV.

Statistical mechanics enables us to evaluate not only the mean values of quantities but also how their actual values fluctuate about the mean. To be more specific, it enables us to ascertain the probability of observing a given difference n from the mean N_0 of the number of molecules in any small volume. Obviously, there is an equal chance of n being positive or negative since N_0 is the average. In addition, small values of n should be more likely than large ones. These two features can be seen in the complete results shown in Fig. 3.11. The values along the vertical axis give the probabilities $P(n)$ of finding any given value of n, plotted horizontally; more precisely, the area $P(n) \, dn$ gives the probability of finding n in the range n to $n + dn$. The bell-shaped curve is called a *Gaussian curve* and is characterized by one parameter only—w, which defines its width. If w is small, the curve is narrow and the fluctuations from the mean are minute. If w is large, the curve is wide, indicating that large fluctuations are quite likely.

The probability when $n = w$ itself is always about 60% of the maximum probability at zero (i.e. 0.24/0.40 in the figure). In fact,

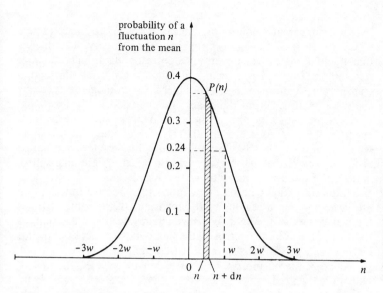

probability of a
fluctuation n
from the mean

$P(n)$

0.4

0.3

0.24

0.2

0.1

$-3w$ $-2w$ $-w$ w $2w$ $3w$

0

n $n + dn$

n

Fig. **3.11** Fluctuations in the number of molecules contained in a given volume of a gas. The probability $P(n)$ is plotted vertically and the *deviation* n from the mean number of molecules N_0 is plotted horizontally. To use the curve, note that the probability that n lies in any range between n and $n + dn$ is given by $P(n)\,dn$, the shaded area. The *width* of the curve is defined by a parameter w equal to $\sqrt{N_0}/2$.

w^2 represents the mean value of the square of n, $\langle n^2 \rangle$, so that w is the root mean square fluctuation.

Theory shows that the value of w as a function of the mean number of molecules N_0 is

$$w = \tfrac{1}{2}\sqrt{N_0}$$

The physically important quantity is the ratio of the root mean square fluctuation w to the mean number of molecules:

$$w/N_0 = \frac{1}{2\sqrt{N_0}}$$

because this tells us how large we can expect the fluctuations to be in relation to the total number in the small volume. We now give an example of what can be deduced from this result together with the Gaussian curve.

The area under the curve between $n = -2.58w$ and $+2.58w$ represents 99% of the total: this means that there is a 99% chance that n lies between these values. Thus the ratio of the actual number of molecules in the small volume Δv to the average value N_0 lies in the range $1 \pm 1.29/\sqrt{N_0}$. If N_0 is large, this is a very small range indeed. For instance, in 1 mm^3 of a gas under normal temperature and pressure, N_0 is of the order of 10^{16}. Our expression shows that the relative fluctuations in this number are normally less than 10^{-8} or less than one-millionth of one per cent. *These fluctuations exist but are imperceptible*: when considering volumes of around 1 mm^3 a gas appears homogeneous, as we can readily see from observation.

However, the fluctuations become more perceptible as N_0 becomes smaller. In a volume of $0.5\ \mu\text{m}^3$, the limits of the relative fluctuations we expect to observe with the same probability reach $1/1000$. This is still small but there is, however, one phenomenon that is affected by this scale of fluctuation: the **scattering of light** by the gas.

In a transparent medium, the density is related to the refractive index, μ—in fact, $(\mu - 1)$ is proportional to density. This means that fluctuations in density produce corresponding fluctuations in refraction. Now when a beam of light is projected through a transparent medium which is not perfectly homogeneous, some energy is removed from the beam and scattered in all directions: this is a prediction of optical theories confirmed in practice. This effect enables us to see a small fragment of glass immersed in water. It also explains how small particles that are too small to be visible can nevertheless be observed in a beam of light, like smoke in the air or impurities in cloudy water (Fig. 3.12).

Scattering of light needs fluctuations in refractive index which are large enough in regions with dimensions of the same order as the wavelength of the light. We have already seen in a gas that, as the size of the volume decreases, the fluctuations in density become larger. We should therefore expect that light of short wavelength would be scattered more than that of longer wavelength: that is, blue more than red.

And for that reason . . . the sky is blue. It does not emit light itself but scatters light out of the sun's rays towards us. If the earth were in a vacuum, the sky would appear black as it clearly did to the cosmonauts on the moon (Fig. 3.13)—but because we have an atmosphere which scatters blue light more than red, the white light from the sun gives rise to an apparently blue sky. Thus, a mere observation of the sky confirms that the air is non-uniform on a scale far smaller than is available to our eyes even with the aid of instruments.

We can go even further. The scattering of light not only demonstrates the existence of molecules but it also enables us to calculate their sizes. For that to be done, it is not enough merely to observe scattering in the atmosphere. It is necessary to measure the fraction of light energy scattered by a gas out of an incident beam

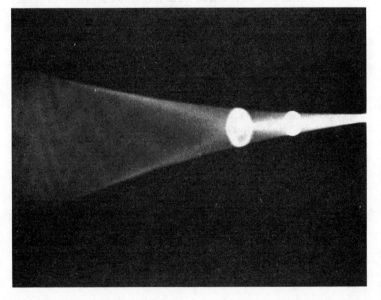

Fig. 3.12 A beam of light projected through air laden with tobacco smoke is made visible by scattering from the small particles of the smoke. The bright elliptical shapes are due to the passage of the light through two fluorescent pieces of glass. (M. Cagnet, Institut d'Optique, Orsay.)

Fig. 3.13 The sky as seen from the surface of the moon is completely black even in direct sunlight. The only visible object is the illuminated portion of the earth's surface. (USIS.)

of radiation with a well-defined wavelength. This fraction can be shown theoretically to be a function of the magnitude of density fluctuations and yields a method of measuring Avogadro's number. (Since fluctuations depend on the number of molecules in a given volume it is clear that they will be related to the number of molecules in a mole.) It is true that the measurements are difficult so that the accuracy of the results is not very high, but there is no doubt that the values for Avogadro's number yielded by this method are in good agreement with those from other measurements based on entirely different principles—and that, as we have pointed out before, cannot be the result of chance.

Is it not both surprising and stimulating that humans, by using their imagination and powers of reasoning, can look at the blue light coming from a clear sky and deduce that in one litre of air there are 27 million million million molecules?

4 Crystals

We now pass directly from one extreme to the other—from a completely disordered state of matter to one that is perfectly ordered, the crystalline state. We do this for two reasons: firstly, this is the only other case in which the atomic models are known exactly, even if they are sometimes very complex; secondly, it is also the only other case where the model allows us to make simple calculations.

The main feature of a crystalline solid has been shown in Chapter 2 to be the abrupt transition which accompanies its formation due to the unavoidable discontinuity that must occur between disordered and ordered states: for instance, in condensation from the vapour, solidification of the liquid, crystallization of a glass, precipitation out of a supersaturated solution, and so on.

One point must first be cleared up. Whereas in Chapter 2 we used the term *crystalline solid*, in this one we shall talk of a *crystal*. The relationship between the two will be studied in more detail in Chapter 6. For the moment we simply say that crystals are the elementary components of a crystalline solid. Occasionally, a crystalline solid does consist of just one single crystal, sometimes quite large as in natural rocks and gemstones. However, this is comparatively rare and a normal crystalline solid, such as an ordinary piece of metal or rock, consists of a collection of crystals of all shapes and sizes compacted together. It seems logical, therefore, to begin by describing the structure of a single crystal.

There is one final preliminary point to be clarified. In Chapter 2 we discussed the changes of state from disorder to order. To simplify the explanations, we took a particular case in which the fundamental 'particle' in every state of matter was the same molecule, and a crystal was defined as an ordered arrangement of molecules. There are, however, crystals in which the particles are not molecules but atoms or ions, even though the gaseous form of the same substance contains molecules. The definition of a crystal must therefore be made more general: *a crystal is an ordered arrangement of atoms or, more precisely, of atomic cores* (that is, nuclei and inner electron shells). These atoms or atomic cores can sometimes be grouped into molecules having the same structure in the solid as in the liquid or gaseous states, but they can *also* be bound in ways that are different in the solid form from that in the disordered states. As examples, metallic sodium is composed of Na^+ ions while the vapour consists of Na atoms and a few Na_2 molecules; a crystal of rock-salt has Na^+ and Cl^- ions alternating throughout the structure but the vapour consists of Na and Cl atoms and a few NaCl molecules.

We shall return later to the types of bond which occur between the atoms in a crystal where there are no molecules. They all concern only the states of the outer electrons and are of the same type as that between the atoms of a molecule. This interatomic binding is certainly very important because with it we can classify crystals using physical properties. Our first concern, however, is with a *geometrical* description of crystals and for this we only need to know the types of atoms and the positions of their nuclei, ignoring for the moment the nature of the bonds between them.

The atomic structure of crystals

The unit cell and the elementary motif

The order which is characteristic of a crystal is caused by *the exact periodic repetition of a basic pattern of atoms called a* **motif**. To appreciate exactly what is implied by that statement, we look first at some examples on a macroscopic scale. In the mosaic border illustrated in Fig. 4.1(a), the *motif* is a pattern which is repeated along a single axis at regular intervals called the *period, a*. In two dimensions, as in the carpet of Fig. 4.1(b), the *motif* is repeated along two directions with periods *a* and *b*. The whole arrangement here can be considered as the juxtaposition of parallelograms all with sides $\vec{a}$ and $\vec{b}$ and with identical contents (the motif).

Similarly, a crystal in three-dimensional space consists of a repeated basic pattern of atoms with a periodicity in three directions having three periods *a, b, c*. This is equivalent to saying that the crystal is a juxtaposition of **unit cells**, each an identical parallelepiped of sides $\vec{a}, \vec{b}, \vec{c}$, and each containing the same repeated **motif** of atoms.

Any one atom of the motif is repeated in every cell and the whole collection of these equivalent atoms forms the 'nodes' of a network called the **crystal lattice**. Figure 4.2 shows that such a lattice can be considered as a set of juxtaposed unit cells of sides $\vec{a}, \vec{b}, \vec{c}$, with a node at every corner.

If the unit cell should contain only one atom, the crystal is said to be *simple* and consists of a collection of identical atoms at the nodes of the crystal lattice. Any crystal, however complex, consists of superposed simple crystals each with the same lattice but with atoms which can be different and which are displaced from each other by certain vectors.

Each atom in a crystal has an equivalent atom in every unit cell and the surroundings of every one are absolutely identical: in a crystal of infinite extent there is no distinction between any of them. Such an atomic model clearly exhibits the *long-range order* characteristic of a crystal.

This model has, then, been constructed from a *lattice* and a *motif*. It is essential to realize that although the lattice was related originally to a certain cell, it can in fact be described in terms of different ones. The lattice is defined simply by the collection of

(a)

(b)

Fig. 4.1 (a) Mosaic border in black and white marble: the pattern is repeated periodically giving a 'one-dimensional network' of period *a*. (Decoration in a Cairo mosque: J. Bourgoin, *Précis de l'art arabe*, Leroux, Paris.) (b) Italian carpet of the 16th century, a two-dimensional network of periods *a* and *b*. (R. Viollet.)

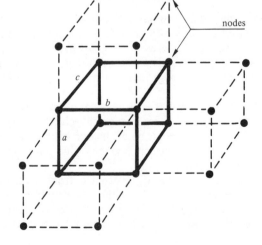

Fig. 4.2 A crystal lattice. The *unit cell* is the parallelepiped with sides *a*, *b*, *c*. The cells placed side by side fill the whole of space: three cells adjacent to the central one are shown. Each corner of a cell is a *node* and the whole array of nodes forms the triply periodic *lattice*. Each unit cell contains the same *atomic motif*, although this is not represented in the figure.

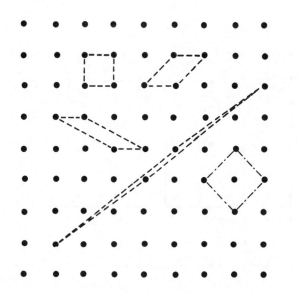

Fig. 4.3 A lattice of nodes in two dimensions. It is easy to check that each unit cell outlined in *dashed* lines can cover all the nodes of the lattice by juxtaposing an infinite number of them. The four examples are all 'primitive' cells having the same area and containing one elementary motif. The cell outlined in *dots and dashes* is not primitive but contains two units of the motif.

nodes and not by any particular cell: indeed, the same lattice can be produced by an infinite number of triplets of three different vectors. This is shown for two dimensions in Fig. 4.3 but it is equally true for a crystal in three dimensions, although the drawing would be much less clear.

The following question then arises: why, if that is the case, is one cell chosen rather than another in the same lattice? The answer is that it is simply a matter of convenience. In general, there is one cell which appears to be the simplest because it exhibits most clearly the symmetry properties of the crystal. This is the case with the lattice of Fig. 4.3, where the small square unit cell is easier to describe and use than any of those in the shape of a parallelogram. However, crystallographers do sometimes use more than one cell to represent the same lattice because the same one is not always best adapted to every problem.

Net-planes or lattice planes

In order to simplify the description of a crystal lattice and at the same time to bring out some of its essential features, we can consider the whole three-dimensional collection of nodes as made up from sub-groups of one or two dimensions. Thus, one way of looking at a crystal lattice is as a pile of *two-dimensional planes of nodes*. Each individual node has the same environment, from which it follows that all the planes must be identical, parallel and regularly spaced from each other. Such a collection is called a **family of net-planes** or a **family of lattice planes**.

A given three-dimensional lattice can be decomposed into sets of net-planes in an infinite number of ways. Figures 4.4(a) and 4.4(b) illustrate from two different viewpoints how some of these sets or families can be chosen in a cubic lattice. As a familiar analogy of this same principle in two dimensions, we may take the example of a plantation consisting of straight rows of equally-spaced trees.

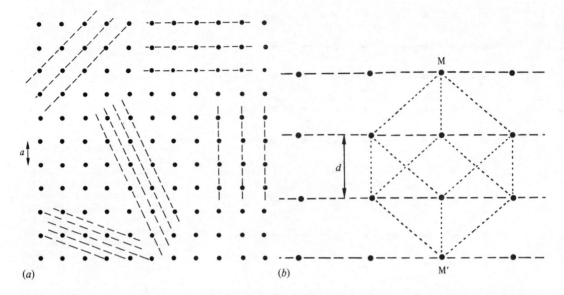

Fig. 4.4 A lattice with a cubic unit cell. (a) All the nodes are projected here on to the face of the unit cells, giving a square lattice of period *a*. The dashed lines represent several sets of lattice planes normal to the plane of the diagram. They appear in cross-section merely as lines. Notice that the wider the spacing of the planes, the more dense are the nodes in the planes. (b) Here, a cubic unit cell is shown from a different angle. The page is parallel to the diagonal MM′ and the planes of nodes are projected on to the page. The spacing of the lattice planes shown, *d*, is one-third of the length of the diagonal.

Anybody walking through it and looking in various directions will see many different sets of rows each with different distances separating them.

In simple crystals (one atom per unit cell), the lattice planes are themselves planes of atoms. In crystals with several atoms per unit cell, there is a family of lattice planes corresponding to each atom (Fig. 4.5). The whole crystal can thus be described as formed from layers of atoms, each type of atom having its family of lattice planes with the same orientation and the same spacing *d*.

A crystal is thus foliated like the leaves of a book. However, whereas a book is formed from a unique set of pages, a crystal comprises an infinite number of families of planes which intersect each other. Of the multitude of families, some are more important because they are connected with macroscopic properties of the crystal. These are the families in which the effect of the 'foliation' is most marked: in other words, those in which the net-planes are most widely spaced and therefore have the greatest number of nodes or atoms per unit area. (As can be seen from Fig. 4.4(a), these two quantities are connected: the reason is that the number of

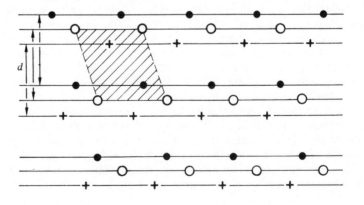

Fig. 4.5 Representation of a crystal with three different atoms (●, O, +) per unit cell. Atoms of the same type occur in lattice planes with the same direction and the same spacing *d*.

(a)

(b)

nodes per unit area in the *plane* divided by the distance apart of the planes is equal to the number of nodes per unit *volume*, and this is clearly the same for all families.)

It is well known that a naturally-occurring single crystal is easily recognized because its outer faces are perfectly plane (Fig. 4.6). In a given type of crystal, the angles between the faces have quite definite values: 90° in rock-salt (NaCl), 120° in prisms of quartz, and so on. Such facts are the basis of the macroscopic study of crystals and are easily explained: *the plane faces of the natural crystals are parallel to certain of the families of lattice planes.* The angles are fixed because they are determined by the geometry of the unit cell. For example, rock-salt has a cubic lattice and its natural faces are parallel to those of the cubic unit cell: in other words, parallel to one of the families of lattice planes with the greatest spacing *d*. Quartz has a unit cell whose shape is that of a prism with a hexagonal cross-section or base: the faces of the prisms of natural rock-crystal are parallel to the lateral faces of the unit cell.

We may add that the existence of a periodic arrangement of 'molecules' in crystals just to explain the regularity of their external shapes had been imagined by René-Just Haüy as much as 150 years before the discovery of the structure on an atomic scale that we are describing. As an example, he showed how the complex shapes of specimens of calcite could be built up from a set of equal small rhombohedra (Fig. 4.7).

Under the action of a sharp blow, an ordinary solid breaks into pieces with irregular shapes, whereas certain crystals *cleave*: the surfaces revealed by fracture are flat planes which form well-defined

Fig. 4.6 (a) A group of quartz crystals, hexagonal prisms terminated by a pyramid. (Photographic library of the Palais de la Découverte.) (b) A group of rock-salt crystals with faces parallel to those of a cube. (N. Bariand, Crystallographic Laboratory, University of P. and M. Curie, Paris.)

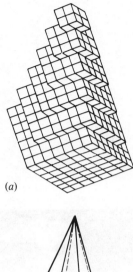

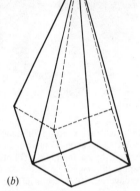

(a)

(b)

Fig. 4.7 The shape of the natural crystal in (b) was explained by R.-J. Haüy in terms of an assembly of identical crystalline units as in (a).

angles with each other. Such 'cleavage planes' appearing in the fractured crystal are also parallel to the net-planes of the lattice. Mica exhibits only one cleavage plane and, as is well known, it can very easily be split into thin sheets (Fig. 4.8). Its atomic structure confirms that there is only one family of lattice planes along which the effects of foliation would be very marked.

In some cases, an electron microscope allows us to 'see' a family of lattice planes as in Fig. 4.9. Knowing the magnification of the instrument, pictures like this enable us to measure the spacing directly.

Net-rows or lattice rows: crystal axes

Crystal lattices can also be broken down in another way, this time into one-dimensional sub-groups. These are identical and parallel *rows*, consisting of lines of nodes spaced at regular intervals. Their arrangement is an ordered one because they originate from the nodes of a lattice plane. Just as with the families of net-planes, it is possible to choose families of lattice rows in an infinite number of ways. However, the important families will be those containing the greatest number of nodes per unit length—and thus will be those that are furthest apart from each other.

A row is the intersection of two lattice planes. So a crystal with natural faces that are parallel to important planes has *edges* which are the intersection of the faces and are thus parallel to important *rows*. These are often called the *axes* of the crystal.

A crystal model can be viewed in such a way that a family of rows is looked at end-on. If they are important rows that are

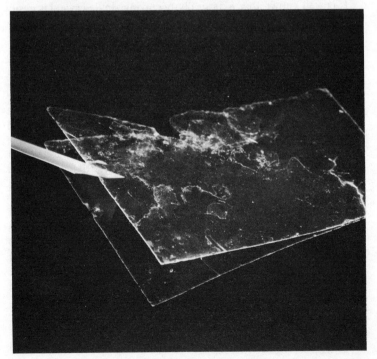

Fig. 4.8 A thin sheet of mica is cleaved in two with the aid of a piece of stiff card. (A. Choisnet, Hachette.)

well-spaced from each other, they often mark off empty channels through the crystal. This can be seen in the photograph of a model of a NaCl crystal shown in Fig. 4.10.

The existence of layer structures in crystals or of channels bounded by lattice rows can be demonstrated by experiments on 'channelling'. High energy α-particles (5 MeV) can pass right through a crystalline sheet if they are projected exactly along the gaps between well-spaced lattice planes or along wide channels between rows. However, when deviated by only a few degrees from those easy directions, they are completely stopped. A point source of α-particles is placed on a single crystal of gold in the shape of a wafer with a thickness of 12 μm in a direction normal to the diagonal of the cubic unit cell (the view in Fig. 4.4(b)). Only those particles which travel parallel to a face of the cube can pass through the crystal and produce an image on a sensitive plate (Fig. 4.11). An equilateral triangle formed by lines parallel to the three faces of the cubic unit cell meeting at a corner (Fig. 4.4(b)) is delineated on the plate.

Some crystals exist with empty channels of particularly large cross-section (diameter greater than 0.5 nm) parallel to one of their axes. Molecules with diameters smaller than 0.5 nm can pass right through the crystals while those with larger diameters cannot. Thus a crystal such as zeolite can act as a molecular 'sieve'—it allows straight-chain aliphatic organic molecules to pass while it stops their isomers with branched chains which make their molecules more bulky. This is a technique used in the oil industry.

Anisotropy and symmetries in crystals

Observation of a crystal shows that it is *anisotropic*. This means that any physical properties depending on direction vary with the orientation of that direction in the crystal. *Anisotropy is a*

10 nm

Fig. 4.9 Lattice planes viewed edge-on as revealed by an electron microscope. Using the known magnification of the instrument, the spacing of the planes can be calculated as 0.7 nm. (C. Collex, Laboratory of the Physics of Solids, Orsay.)

Fig. 4.10 Photograph of a crystal model of NaCl emphasizing the channels parallel to the edges of the cubic unit cell. (Photographic library of the Palais de la Découverte.)

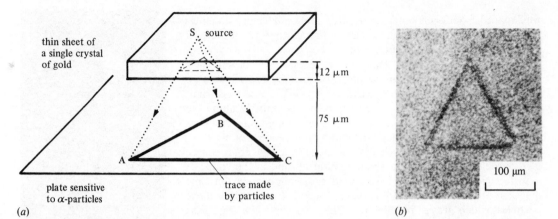

(a)

(b)

Fig. 4.11 An experiment on channelling. (a) The experimental arrangement. Particles passing right through the crystal are confined to planes SAB, SBC, SCA, which are parallel to the faces of the cubic unit cell, the diagonal of which is vertical. (b) A trace made by the α-particles on a sensitive plate of cellulose nitrate treated with caustic soda. (Y. Quere, CEA, Fontenay-aux-Roses.)

fundamental characteristic of crystals. Some examples have already been given: for instance, certain crystals can be easily cleaved parallel to some planes and not to others; when a synthetic crystal is produced (by very slow evaporation of a saturated solution, for example), it grows with plane faces making definite angles with each other and not with the shape of a sphere; the refractive index of light varies according to the direction of the light in the crystal; and so on.

In fact, long before there was any idea of the structure of crystals on an atomic scale, the attention of naturalists was first drawn to some remarkable minerals by such anisotropy and in particular by the regularity of the external shapes already mentioned.

Although the properties of a crystal vary with direction, it is found that there are a certain number of different directions in which the properties are identical. Such directions are said to be *equivalent.* The sets of equivalent directions are related in symmetrical ways, as we shall see: the crystal is said to have *symmetry properties.*

Symmetry operations allow us to find the directions which are equivalent to a given direction. Some examples will make this clear. A NaCl crystal, after a rotation around one of its edges by either 90° or 180° or 270° has the same properties in all directions as it had originally. Since there are four equivalent positions at 90° to each other, the crystal is said to have a *rotation axis of symmetry of order four* or a *four-fold rotation axis.* Figure 4.12(a) shows that, after a 90° rotation about an edge, the cubic unit cell can be superposed on its original position by a simple translation. Note, however, that an actual specimen of a real crystal does not behave like this (Fig. 4.12(b)). The reason for this is that the specimen is made up of an arbitrary number of unit cells and the whole is not geometrically similar to one of them. The external shape of a crystal is not one of the *intrinsic properties* of the substance and it is only these we are interested in. A quartz crystal has the shape of a hexagonal prism surmounted by a pyramid. When the crystal is turned through 60° (= $2\pi/6$ radians) its faces end up with that orientation with which it started, so it looks as if it has a six-fold

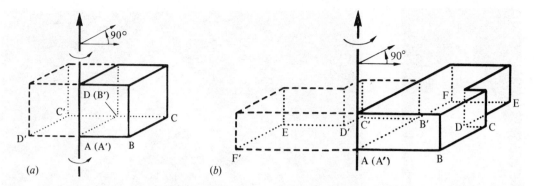

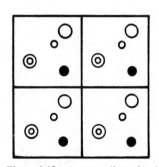

rotation axis. However, a more careful inspection shows that there are small facets at the junctions between the faces of the prism and the pyramid which are well-developed on only three of the edges. So if the crystal is to be turned into an *exactly* equivalent position, it needs a turn of 120° (= $2\pi/3$ radians): quartz has thus a three-fold rotation axis.

The symmetry operations of a crystal, determined by observation, and the way they are associated with each other are studies which have led mineralogists to a systematic grouping of crystals into *seven crystal systems* each subdivided into classes.

Relation between the atomic structure and the symmetries of a crystal

We have been approaching the structure of crystals from two different points of view: firstly, on the atomic scale, based on the idea of a **motif** of atoms which is regularly repeated by translations allowed by the **lattice**; secondly, on the macroscopic scale, based on symmetry properties. It is obvious that these two viewpoints must be compatible because they are aspects of the same reality: a strictly ordered form of matter. In fact, it turns out that a classification of crystals according to the geometrical character of the unit cell (ranging from the cube through to the ordinary parallelepiped) coincides with the seven symmetry systems of the mineralogist.

However, to make things quite clear there is one false idea that must be avoided. It is one that is not obviously excluded by the description of a crystal as being generated by a motif and a lattice. It is possible to imagine a group of atoms as a motif which has no symmetry relation at all to the unit cell which contains it. The two-dimensional example in Fig. 4.13 shows such a case. The motif has none of the symmetry properties of the unit cell—and therefore neither does the crystal as a whole. Such a scheme, however, does not correspond to reality, because in a solid the atoms are packed tightly against each other—they have short bonds, not only with those atoms in the same cell but also with those in immediately adjacent cells. It follows that the shape of the cell is not independent of the arrangement of atoms in the motif: the cell and the motif have certain minimum symmetry properties in common. Thus, *a crystal system is characterized both by the shape of the unit cell and by a minimum number of elements of symmetry*. In the

Fig. 4.12 (a) A symmetry operation: rotation by 90° about a vertical axis ABCD to A′B′C′D′. The new position of the cube can be reached from the original one by a simple translation from right to left. (b) This is not true for a crystal specimen of the cubic system with an irregular shape. However, the *physical properties* of the crystal after rotation are the same as those before, in all directions.

Fig. 4.13 A two-dimensional 'crystal', with an arbitrary motif of atoms in a square cell. The crystal does not have the symmetry properties of the unit cell.

(a)

(b)

(c)

|— 10 μm —|

(d)

(e)

(f)

(g)

(h)

cubic system, for instance, we must have: (a) a cubic unit cell and (b) four 3-fold rotation axes in the crystal along the diagonals of the cube.

We see then that there are two complementary aspects of order in crystals: repetition with triple periodicity and symmetry properties of the whole. We also see that there must be compatibility between the two and we indicate two more examples of this:

1. On p. 64, we noted the existence of a 4-fold rotation axis in a crystal of NaCl. That was a particular example. In general, there are *n*-fold rotation axes in crystals which correspond to an operation of rotation through $2\pi/n$ radians. *The number n must be an integer* so that a finite number of successive operations brings the crystal back to its original position. When translational periodicity exists, as in a crystal, it can be shown that *only 2-, 3-, 4- and 6-fold rotation axes can occur*. In particular, even if an isolated object can have pentagonal or 5-fold symmetry (as in many flowers), it is never encountered in crystals (Fig. 4.14). It is not possible to fill an area by indefinitely repeating regular pentagons which touch each other as it is with triangles, squares and hexagons.

2. In Fig. 4.15(a) there is a motif of two identical 'atoms' (circles), symmetrically placed with respect to XX', which is

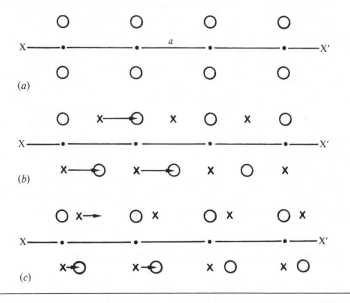

(a)

(b)

(c)

Fig. 4.15 (a) A one-dimensional 'crystal' of period *a*, symmetrical with respect to XX'—that is, reflection in XX' will produce an identical pattern. (b) *Symmetry with glide equal to half the period a/2*. The pattern of circles when reflected in XX' gives the pattern of crosses. To make this identical with the original, there must be a translation of *a/2* as shown. Figure 4.1(a) was an example of glide symmetry of this sort. (c) If this pattern of circles is reflected in XX' to give the pattern of crosses, there is no way of superposing the two by translation only.

Fig. 4.14 Axial or rotational symmetries in crystals: (a) 2-fold axis: calcite, (b) 3-fold axis: internal striations in a tourmaline crystal (N. Bariand), (c) 4-fold axis: MgO crystals under an electron microscope and (d) 6-fold axis: snowflake.

5-fold rotation symmetry in plants and animals without periodic structures: (e) bell-flower (campanula) and (f) starfish.

Symmetries in extended patterns: (g) identical equilateral triangles alone can form a periodic two-dimensional pattern (design in marble from a Cairo mosque) and (h) identical regular pentagons cannot do the same: hence the variations in a ceramic pattern (Isfahan).

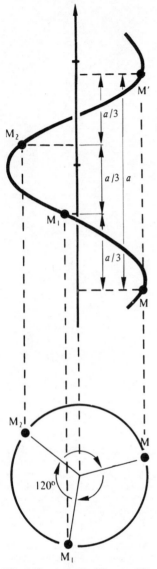

Fig. 4.16 A motif of atoms with a *3-fold screw axis*—that is, the atom M is displaced to M_1 by a rotation of 120° $(2\pi/3)$ coupled with a translation of $a/3$. M_1 is displaced to M_2 and M_2 to M' by the same operation. M' is the equivalent of M and is related to it through a simple translation of period a.

repeated along that axis at intervals of a. The whole pattern is identical with its mirror image or reflection in XX'. Now consider the motif of Fig. 4.15(b), where the lower of the two atoms (circles) has been displaced to the right by $a/2$. This time, the same repetition period a no longer makes XX' a simple axis of symmetry. Instead, reflection in XX' yields the pattern of crosses which can be brought into coincidence with the original circle pattern by a translation of $a/2$. This is called *glide symmetry* as opposed to the simple symmetry of Fig. 4.15(a). Figure 4.15(c) shows that the repeated motif does not yield glide symmetry unless the lower atom is displaced by *exactly* $a/2$. Note particularly that the two forms of symmetry in Figs 4.15(a) and 4.15(b) have the same macroscopic effect for the crystal as a whole: the symmetry with respect to XX'. It is for the isolated motif that the two operations differ.

Another set of symmetry operations is also made possible by the periodicity in crystals. These are *screw axes*: a combination of rotation about an axis and translation along it equal to a simple fraction of the period (1/2, 1/3, 1/4 or 1/6). An example is shown in Fig. 4.16. Such screw axes may have helices that are right-handed or left-handed. It is because of this that quartz crystals, which do contain screw axes, are of two kinds: one laevo-rotatory and the other dextro-rotatory (rotation of the plane of polarized light passing through the crystal).

Without reference to experiment, mathematicians have deduced quite logically all the possible combinations of symmetry elements and translations which are intrinsically distinct from each other. There are found to be 230 different combinations called *space groups* and all ordered atomic structures must inevitably belong to one of these groups. It is true that the allocation of a structure to a particular space group only provides a sort of skeleton that the physicist must flesh out to establish the complete structure—but we can see how much this guide should be a great help: it never breaks down since it is impossible for a real structure to be unclassifiable. Some space groups are observed to be very common and others very rare: there are even some for which there is no known structure, but they are becoming fewer as the number of established structures increases. We conjecture from this that nature has probably used all the arrangements of atoms permitted by geometry.

Determination of crystal structures

By what methods do we succeed in building up models of crystal structures? The results are of crucial importance to our knowledge of the structure of matter, but a detailed description of the techniques used is outside the scope of this book. Nevertheless, we must at least give a clear idea of what can be achieved by current experimental methods.

The starting point is the known chemical composition of the crystal. This gives us information about the number and type of atoms in the motif since it obviously has the same composition as the whole substance. For example, in barium sulphate the motif must consist of the following atoms: one Ba, one S, four Os, or *an integral multiple* of this group. In addition, we know the radii of the hard spheres that can represent ions (see Tables 1.1 and 1.3) and the distances between covalently-linked atoms (see Table 1.4) as well as the space occupied by them (p. 9). Finally, the symmetry of the structure is determined by the macroscopic properties. Those are the data: the problem is then to find the unit cell of the crystal and the positions of the atoms within that cell.

The structure of NaCl

We first give a very simple, but historically important, example— one in which the structure could be guessed but which at the same time serves as the starting point for modern crystallographic methods. At the beginning of the century, an English mineralogist named Pope had conceived a model for the NaCl structure from two starting points: the cubic symmetry of natural crystals and the idea that Na^+ and Cl^- ions were packed alternately so that electrostatic forces could ensure the cohesion of the whole assembly. Pope's model has been well confirmed as the now classic arrangement shown in Fig. 4.17: one cubic unit cell containing a motif of eight ions, four Cl^-s and four Na^+s. We pass from this qualitative representation to a quantitative one by determining the length of the side of the cube, a, as follows. Since ordered matter is homogeneous, the density of the cell itself is the same as that which is determined from a macroscopic crystal, denoted by ρ. If the molar masses of Na and Cl are respectively M_{Na} and M_{Cl}, then the masses of the *ions* are each M_{Na}/N and M_{Cl}/N, where N is Avogadro's number. The total mass in the unit cell is thus $4(M_{Na}/N + M_{Cl}/N)$ and since its volume is a^3, we have

$$\rho = 4(M_{Na} + M_{Cl})/a^3 N$$

from which the value of a can be calculated. The macroscopic density is found experimentally to be 2.163×10^3 kg m^{-3} and this gives $a = 0.5628$ nm.

This value is consistent with the radii of the Na^+ and Cl^- ions of 0.100 nm and 0.181 nm respectively given in Table 1.1. The sum, 0.281 nm, should give the Na–Cl distance AB in Fig. 4.17 as $a/2 = 0.5628/2 = 0.2814$, which it clearly does very closely.

The value of a for NaCl, called its lattice parameter, is interesting because it gives some idea of the sizes of crystalline unit cells in general. They are mostly between 0.4 and 2 nm, although for some crystals with very large molecules they may be as large as 5 nm.

The use of electron microscopes

Atomic diameters have the same order of magnitude as the resolving powers of the best electron microscopes. Indeed, in certain exceptional cases, images have been obtained which can be

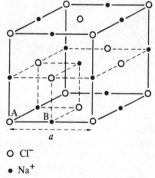

○ Cl^-

● Na^+

Fig. 4.17 The unit cell of a NaCl crystal (rock-salt). It can be described in two ways: as a cube of side 0.5628 nm containing a motif of four Cl^- and four Na^+ ions, or as a *face-centred cube* with the same parameter and a motif of one Cl^- ion at A and one Na^+ ion at B. *Note*: while the ions are at the nodes of a cube of side 0.2814 nm, this is *not* a unit cell since the nodes are filled by two sorts of ions.

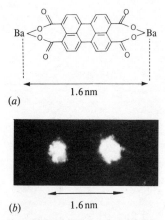

(a)

1.6 nm

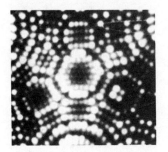

(b)   1.6 nm

Fig. 4.18 (a) Molecular structure of barium perylene tetracarboxylate. (b) Image of the same molecule obtained with a 3 MeV electron microscope. Two spots appear at a distance apart of 1.6 nm, the same as that between the two heavy atoms, barium, predicted by the molecular structure. The shape of the spots is not significant. (Dorignac and Jouffrey, Electron Optics Laboratory, Toulouse.)

Fig. 4.19 Image in a field emission microscope of a platinum crystal in the shape of a sharp point. Its surface is formed by a number of different lattice planes whose atomic structure is revealed. (J. Gallot, University of Rouen.)

interpreted as views of individual atoms (Fig. 4.18). Would it not be possible, then, to use this instrument to obtain a direct image of the atomic structure of crystals? The difficulty is that, even if the theoretical resolving power is good enough, we still have to prepare the specimen in a form adequate for the instrument to allow us to 'see' the structure. It is in fact only in rare cases that an image of a crystal lattice can be said to have been obtained. Nevertheless, exceptional as these images are, they obviously have such a considerable interest that we give several examples of them.

The *field emission microscope* is an instrument capable of yielding an image with a useful magnification of up to 10 million from the surface of the hemispherical point of a very sharp needle. With such a magnification, 1 nm on the object is represented by 1 cm in the image. The point of the needle consists of a single crystal with a radius of about 10 nm. On this scale, its surface is not smooth because the successive lattice planes form steps and, in addition, even over a single lattice plane the individual atoms cover the surface with raised bumps. The blurred spots that can be seen in the photograph (Fig. 4.19) correspond to the positions of the individual atoms, and it is also possible to recognize the different lattice planes that are exposed on the surface of the point. Figure 4.19 was obtained from a platinum specimen and all the details in it can be completely and *quantitatively* explained in terms of the accepted atomic structure of platinum. Such excellent pictures can only be obtained with a few metals.

With a carefully aligned *electron microscope of the normal type*, pictures have been obtained of the surface lattice planes of certain crystals with exceptionally large molecules having dimensions of the order of 20 nm. A few details can even be seen of the internal structure of the molecules, identical in every case (Fig. 4.20).

There are some other remarkable examples using thin sheets of only about 5 nm in thickness. When these are formed by the stacking of successive atomic layers one on top of another, observation of the sheet in a direction strictly normal to its plane gives images of the individual layers that are superposed without becoming confused. The two-dimensional lattice in the plane parallel to the sheet then appears quite clearly and its parameters can be measured on the photograph using the known magnification of the instrument. If the motif is one molecule, its shape as defined by its chemical formula can sometimes be recognized (Fig. 4.21).

In germanium, the sheet is so cut that the atoms form zig-zag chains perpendicular to its plane surface (the chains are illustrated in Fig. 1.4(c)). In projection, the atoms in each chain are not resolved because their apparent distance apart (0.14 nm) is less than the resolving power of the microscope (about 0.28 nm). However, the rows in the lattice planes parallel to the surface show up very clearly in the photograph and their arrangement agrees exactly with what is known of the structure of germanium (Fig. 4.22). It is thus by no means an exaggeration to say that we have now succeeded in seeing atoms in ordered states of matter. (By contrast, no current techniques will give us images of atoms in *disordered* condensed matter, whether it is a liquid or an amorphous solid.)

X-ray diffraction

The very best experiments giving a *direct image* of a crystal structure are only effective in exceptional cases. However, there is a method which has the merit of being perfectly general and which is capable of revealing the structure of any crystal whatsoever: this is one that uses the diffraction of X-rays.

X-rays are waves of the same nature as light but with a wavelength about a thousand times shorter—of the order of 0.1 nm. When a crystal is irradiated with a fine beam of X-rays, each of the atoms 'scatters' the rays: that is, it becomes itself a weak source of waves propagating in all directions. These reradiated wavelets interfere with each other.

The results of this interference can be most simply explained as follows: the scattering from the whole collection of atoms *cancels out* in all directions *except* in some where there is great reinforcement: these directions are very exactly defined and are quite numerous (Fig. 4.23). The crystal is said to produce *diffracted* rays or beams which arise from two causes: one is the regularity of the atomic lattice and the other is the fact that the wavelength of X-rays is of the same order as interatomic distances.

It is extremely important to understand the following principles. *The **directions** of the diffracted beams are determined by the crystal **lattice**; the relative **intensities** of the different beams are determined by the **motif** of atoms in the unit cell.*

Conversely, the *shape and size of the unit cell* can be deduced from the *directions* of the diffracted beams: this is always possible and is comparatively simple. To determine the *positions* of the atoms within the unit cell, it is necessary: (a) to measure the *intensities* of a very large number of diffracted beams and (b) to use methods of calculation that are somewhat tricky and need a lot of computer time.

The methods of X-ray diffraction do not give an image of the atomic structure but they do allow it to be reconstructed. The

Fig. 4.20 Image in an electron microscope of a protein crystal. The molecular lattice has parameters of the order of 12.5 nm. Some details of the interior of the molecule can just be seen. (T. L. Blundel and I. N. Johnson, *Protein Crystallography*, Academic Press, New York, 1976.)

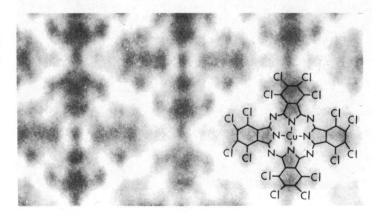

Fig. 4.21 Image in a very high resolution electron microscope of molecules of copper hexadeca-chlorophthalocyanate. The shape of the molecule given by its formula can be recognized as that of the motif of the crystal in the picture. (N. Uyeda, Kyoto.)

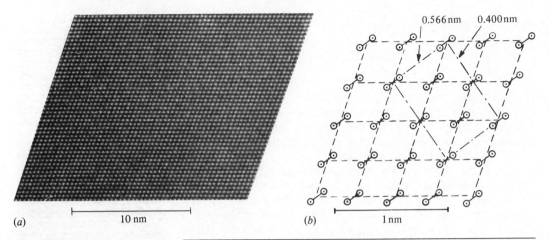

(a) |———   10 nm   ———|

(b) |———   1 nm   ———|

0.566 nm   0.400 nm

Fig. 4.22 Image in an electron microscope of very high resolution (about 0.25 nm) of a germanium crystal. Germanium has the same structure as diamond (Fig. 1.4(b)). The specimen is a thin sheet of thickness 5 nm cut parallel to lattice planes along the vertical lines in Fig. 1.4(b). The atoms form zig-zag chains as in Fig. 1.4(c) which are viewed end-on in the specimen and so appear as in Fig. 4.22(b) here. In this projection, the pairs of atoms are separated by only 0.14 nm and are not resolved. The pairs thus appear as a single white spot in the microscope image of Fig. 4.22(a), which is thus exactly the same as the theoretical pattern. Since the magnification of the instrument is known, the photograph gives directly the absolute value of the lattice parameter of germanium. (A. Bourret, Centre for Nuclear Studies, Grenoble.)

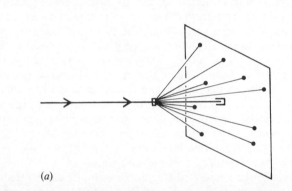

(a)

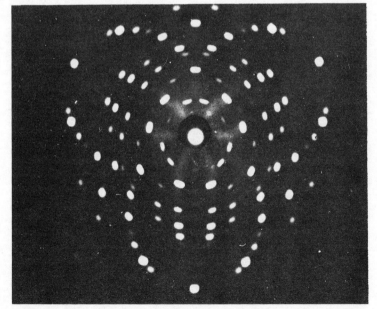

(b)

Fig. 4.23 (a) Experimental arrangement in Laue's experiment. A small fixed crystal is irradiated by a fine beam of X-rays having a continuous range of wavelengths. The *diffracted rays* are recorded on a photographic film normal to the X-ray beam. (b) Example of a Laue pattern given by a crystal of calcium fluoride (CaF_2). (Photographic library of the Palais de la Découverte.)

methods are in principle the same as with the diffraction and interference of visible light in Young's double slit experiment. From measurements on the observed fringes, it is possible to calculate back and find the separation of the slits and theoretically even to determine their individual shapes. The principle of the method was discovered about 65 years ago by Max von Laue and by W. H. and W. L. Bragg. Since then, apparatus has been continually perfected and automated, the theory has been refined, and, most recently and most importantly, crystallographers now have powerful computers at their disposal.

Here is an example of what is possible nowadays. We start by taking a small single crystal with dimensions of a few tenths of a millimetre and place it in an automatic diffractometer (Fig. 4.24(a)) which gives the parameters of the unit cell in about 20 minutes. The apparatus is then required to record the intensities of the different diffracted X-ray beams, a process that involves automatically-controlled rotations of the crystal and of the X-ray detector. A unit cell with dimensions of 0.5 to 1 nm can have several thousand diffracted beams to be measured and the instrument needs to function continuously for about two days to collect all the data. If the unit cell contains less than a few dozen atoms, the crystallographer can normally, after one or two weeks of further work and calculation, determine the structure of the crystal: that is, the positions of the atoms in the unit cell and, if a molecule exists in the crystal, its conformation (in organic substances, for example). An outline of the process is given in Fig. 4.24(b).

However, it is certainly not always as easy as that. In the first place, more and more studies are being carried out on very large molecules in the biochemical field: the amount of work, both experimental and theoretical, is incomparably greater, yet the structure of a protein containing 5000 atoms in its molecule has already been determined in six months. There are also some difficult problems which, for one reason or another, resist routine methods. Nevertheless, it is not an exaggeration to say that today, by and large, the problem of crystal structures has been solved. Using the methods we have described, which are indirect but very reliable, we can 'see' the structure of the crystal or of the molecules on an atomic scale. This is an important development for chemists since current techniques put it all within their reach: not much more than 30 years ago, structure determinations were reserved for a small number of specialists and demanded time that could not be found from the normal course of chemical research.

There is one point that must be emphasized: X-ray diffraction can only yield the structure of molecules because they are arranged periodically in a crystal. When they are disordered, as in a gas or a liquid, X-rays are powerless. It is thus essential to know how to prepare a specimen in the form of a single crystal of sufficient size (about 0.1 mm) and this is sometimes the most tricky process in the investigation. Some substances, and it cannot always be predicted which ones, crystallize badly or not at all (Chapter 8) and the solid remains amorphous.

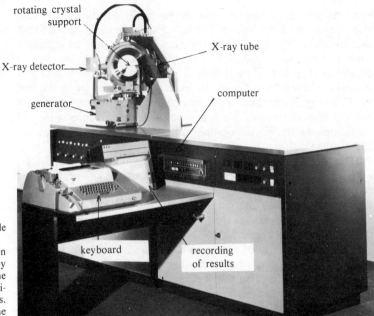

rotating crystal support

X-ray tube

X-ray detector

computer

generator

keyboard

recording of results

Fig. 4.24 (a) Automatic 4-circle diffractometer PW1100 (Philips).

(b) Example of the determination of a crystal structure in 1979 by B. Brachet and C. Brassy at the Crystallographic Laboratory, University of P. and M. Curie, Paris. *Nature of the crystal*: dithine $C_{14}H_{14}O_4N_2S_2$. *Formula*:

(a)

$b = 0.528$ nm $c = 0.987$ nm

$90°$

$96.75°$

$90°$

$a = 1.447$ nm

(1)

0.133 nm

0.176 nm   0.182 nm

0.182 nm   0.177 nm

0.127 nm

(2)

Phase 1: determination of the unit cell and lattice parameters, diagram (1). Time taken: 1 hour. A calculation from the volume of the unit cell and the density of the crystal shows that there are two molecules per unit cell.

Phase 2: recording of intensities of diffracted beams; correction of measurements; treatment of data; checking. Time for setting up and adjustments, 3 days. Computing time, 1 hour 30 minutes.

Phase 3: calculation of crystal structure from values of intensities using a computer program. The first rough shape of the molecule is obtained, diagram (2). Time for setting up and adjustments, 13 days. Computing time, 20 minutes.

Final phase: progressive refinement of results. The difference between observed data and those calculated from the assumed structure is reduced to 4%. The final structure is drawn automatically for stereoscopic viewing. Diagram (3) shows a single view. Time for setting up and adjustments, 13 days. Computing time, 3 hours 20 minutes.

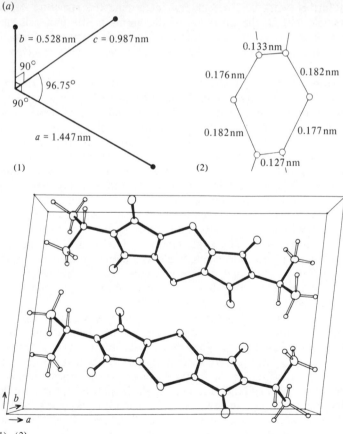

c b

a

(b)   (3)

In Chapter 2, the idea of a crystalline structure was introduced in a rather dogmatic way. We can now claim that atomic models of crystals are based on experimental evidence, especially after the detail we have given of current results in the determination of structures and of the various direct images of crystals that have been obtained. The structures of all crystals have not been determined, but all of them *could* be: from now on, the atomic arrangements in the ordered state of matter hold no mysteries for us.

Bonds between atoms in crystals

Until now we have described the atomic structure of a crystal using a purely geometrical point of view. Yet the very existence of a crystal implies that there is a cohesive force between the atoms and it is the equilibrium between these interatomic forces that determines their positions. For that reason, we shall now consider the crystal from the physical point of view. This is essential if we are to throw light on why certain structures are observed and if we are to predict relationships between structures and physical properties. These, then, are our goals, ambitious and only partly achieved, but we now try and see how far we can progress towards them.

First of all, we shall examine the types of bond between atoms because they enable us to establish a *physical* classification of crystals which proves extremely useful when put alongside the *geometrical* classification of crystallographers. In Chapter 1, we distinguished between the different particles from which crystalline matter is constructed—molecules, ions and atoms. We deal with each of these in turn because they lead to different types of bond.

Molecular crystals

The first category we consider is that of molecular crystals, formed by a collection of identical molecules packed densely together. We recall that a molecule is a group of atoms in a stable combination so that it keeps its identity intact when the material passes from one state to another.

The *internal* structure of the molecule (i.e. the bonds between its atoms) constitute one of the fundamental problems of chemistry and do not concern us here: for us, the molecule is a rigid 'particle' which we do not penetrate. Our concern is to explain how and why these particles collect together in a compact fashion in crystals.

The Van der Waals bond
There is a force of *attraction* between two identical molecules that is very weak and decreases very rapidly as their distance apart increases. It is known as the *Van der Waals force*. Now, because the positive charges on the nuclei are exactly balanced by the negative charge of all the electrons, molecules are electrically neutral. Yet the Van der Waals force has an electrostatic origin, which clearly needs an explanation.

We start with the simplest case but one that appears most paradoxical: the interaction between two atoms of an inert gas. The electron shells in these atoms are all full, so that the negative charge density has a spherical symmetry around the positive point charge of the nucleus. It follows from a well-known result in electrostatics that the electric field outside the atom is zero. Such an atom, A, cannot therefore act on the charges in another atom, B, whatever it may be. However, what we have forgotten in this argument is that an electron in an atom is not stationary but possesses kinetic energy.

We shall adopt the naive picture of point electrons revolving around the nucleus since this classical model is sufficient to predict a result very close to that given by quantum mechanics. There are Z electrons, each with charge $-e$, and a nucleus with a total charge $+Ze$, so that each electron can be associated with one of the fixed $+e$ charges of the nucleus. Such a combination forms an *electric dipole* which produces an electric field outside the atom. As the electron moves around the nucleus, the field at any fixed point varies with time. If there is no *privileged* position, there is also no privileged direction for the axis of the dipole and so the field produced at each point would have a zero mean value: which is exactly what we have already said. However, *at any instant* the field is *not* zero and it can deform the atom B a little by pulling its +ve and −ve charges in opposite directions. B thus becomes an electric dipole itself, always attracted to A, and with a certain interaction energy at any instant with the field created by A. A calculation shows that the mean value of this energy over a period of time, denoted by U, is *not* zero[1]. It is found to be proportional to r^{-6}, where r is the distance between the centres of. A and B. It follows that there is an attractive force $F = -\mathrm{d}U/\mathrm{d}r \propto r^{-7}$ between A and B. This is the Van der Waals force.

When the atoms get close to each other, a second and quite different phenomenon occurs: a repulsion at short distances due to Pauli's principle (see p. 15). The orbitals of each atom are saturated and cannot accept another electron from each other. All the electrons of B must thus remain in orbitals of A outside the closed shells and vice versa. It is clear that the energy of the pair A–B must increase very rapidly from the moment the atoms 'touch' each other.

The curve of Fig. 4.25 represents the sum of the two effects, the attractive and the repulsive. The quantity $U(r)$ is the energy needed to achieve a separation r of the atoms A and B starting from a state where they are very far apart. When $U(r)$ is negative, it means that an amount of work $-U(r)$ is needed to separate the atoms by an infinite distance. The value of r for which the work is a maximum (and the curve a minimum) is the equilibrium separation of the two atoms—that for which they are most strongly bound, and for which the force $F = -\mathrm{d}U/\mathrm{d}r$ is zero. This separation is what we have called the diameter d of the atom when it is represented by a

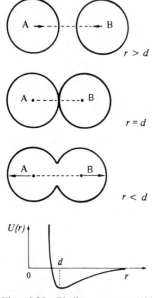

Fig. 4.25 Binding energy $U(r)$ between two inert gas atoms and its variation with separation r. Attraction when $r > d$, repulsion when $r < d$, equilibrium when $r = d$.

---

[1] This is because U depends on the *square* of the mean dipole moment. Even if a quantity itself has a mean value equal to zero, it does not follow that the mean of its square is also zero. Alternating current is another example.

hard sphere. The minimum of the curve is in fact quite flat, and this is why the separation can be made to vary slightly on either side of *d* without the expenditure of much energy: the atomic spheres are not completely hard.

In a multi-atomic molecule, the distribution of +ve and −ve charges is complex and there is no longer any reason why the molecule should not produce an external electric field, even though it is neutral overall. In fact such fields are weak and localized in the immediate neighbourhood of the molecule because the effects of the +ve and −ve charges tend to cancel each other. In other words, the field decreases rapidly with increasing distance from the centre of the molecule. The H_2O molecule, for instance, with the configuration shown in Fig. 4.26 derived from experimental data, behaves like an electric dipole: it is said to be *polar* since the dipole moment is a permanent one (our spherical atoms were *non-polar*). The electric field produced by such a molecule acts on the charges in neighbouring molecules and there is an electrostatic interaction giving rise to a force that is also called a Van der Waals force. However, the mechanism of this one is rather different from that of the one we described in the case of non-polar molecules or atoms with spherical symmetry.

The calculation of the dipole moment of multi-atomic molecules can be carried out irrespective of their complexity. Unfortunately, such a calculation can only be approximate because the exact arrangement of charges is not known. This is because of the deformation of the electron clouds in each atom by chemical bonds and because of the possible transfer of electrons from one atom to another (ionicity—see later on p. 86).

We saw in Chapter 1 (p. 14) that a molecule could be represented by a rigid model whose surface defines a region into which other molecules cannot penetrate. Moreover, we have just seen that the molecules are attracted to each other by Van der Waals forces. In a molecular crystal, therefore, molecules arrange themselves in such a way that they come into contact with the greatest possible energy of cohesion. The resulting arrangement is not always the one that seems the simplest or most likely. It is quite common, for instance, for molecules not to be stacked parallel to each other but to have alternating orientations in a molecular crystal (see Fig. 4.44): calculation of electrostatic energy shows that this arrangement is in fact that predicted from energy considerations.

The hydrogen bond

When a molecule has round its edge an OH or an NH group, the electron of the H atom is transferred to the O or N which are very electronegative. This produces an electric dipole with the electron at one end and the 'naked' positive proton at the other, outside, end. Such a dipole easily deforms (or polarizes) a neighbouring O atom, making it into another dipole and thus attracting it. This is called a *hydrogen bond*. It is responsible for the links between the H_2O molecules in ice and plays a role in a large number of organic crystals. The hydrogen atom produces a link between two atoms,

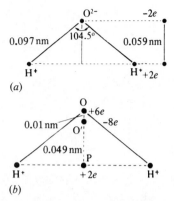

Fig. 4.26 (a) Structure of the polar molecule H_2O giving positions of the nuclei of the O^{2-} and H^+ ions. If the electrons were distributed symmetrically about the O^{2-} nucleus, the dipole moment of H_2O would be $2 \times 0.097 \cos 52.2°$ or 0.12 electron-nm. (b) In fact, experimental results show that it is 0.04 electron-nm, which corresponds to the distribution shown. There is a charge $+2e$ at P, the centroid of the two protons, H. There is a charge $+6e$ at O, the sum of $+8e$ on the nucleus and $-2e$ from the 1s electrons. There is a charge $-8e$ at O′, centroid of the outer shell of electrons. O′ is 0.01 nm from O because of the deformation of the ion by the O–H bonds.

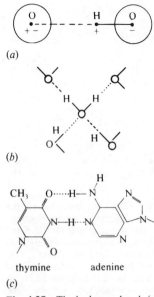

(a)

(b)

(c)

thymine   adenine

Fig. 4.27 The hydrogen bond. (a) The left-hand O atom is deformed by polarization due to the proton H⁺. (b) Hydrogen bonds in ice: each O atom is linked by hydrogen bonds to four neighbouring O atoms. (c) Hydrogen bridges between two bases in neighbouring chains of DNA.

which may be O or N or F, contrary to the normal rules of valency, thus: $O-H\cdots O$. It is again the result of electrostatic action and the H atom is not symmetrically placed between the others (Fig. 4.27).

Characteristics of the Van der Waals bond

If we have been putting a lot of emphasis on the Van der Waals force, it is because it plays a very important role in the structure of a considerable proportion of solids. Although there is nothing mysterious about its origin, it cannot be calculated exactly for complex molecules. We summarize its main features and their consequences as follows.

1. *The binding energy associated with the Van der Waals force is small.* It is only about 0.1 eV (10 kJ mole⁻¹) between the neighbouring molecules in a crystal and is thus between *10 and 100 times smaller than the chemical binding energy between atoms inside a molecule.* The difference is so great that there can be no confusion between intra- and inter-molecular bonds: that is, between those *within* molecules and those *linking* them. This justifies our model of some crystals as 'constructed' from molecules.

2. *The Van der Waals force decreases very rapidly with increasing distance.* It is thus involved only in producing cohesion between molecules 'in contact', in disordered states such as liquids and amorphous solids as well as in crystals. Such forces also exist between gas molecules, but because the distance apart is so large the interaction becomes completely negligible, and justifies the perfect gas model. Under conditions of lower temperatures and higher densities, we have seen that gases begin to deviate from the perfect state and this is because the Van der Waals force becomes significant. However, because the force has a strength that depends essentially on the structure of the particular molecules, the deviations do not occur at the same conditions for every gas but vary from one to another (p. 40).

Ionic crystals

The *ions* which form the 'particles' in this type of crystal were defined in Chapter 1. They are charged particles, positive when the atom has lost exterior electrons and negative when it has captured some. Matter is electrically *neutral* so that the positive and negative charges must compensate each other. Moreover, *the neutrality must occur on as small a scale as possible*, and that is why +ve and −ve ions alternate regularly in ionic crystals. Each ion is then attracted by its nearest neighbours and, although repelled by the next-nearest, these are at a greater distance and so the repulsion is weaker. The well-known model of NaCl in Fig. 4.17 is the simplest example of this alternation of charges in three-dimensional space.

X-ray diffraction methods will not only give the positions of the atomic centres in the unit cell (p. 71) but can also, when used to the full, provide a three-dimensional map of the electron density in the cell. Figure 4.28 shows a one-dimensional section through such a map. It gives the experimental values of electron density on the line of the bond between a Na⁺ ion and a Cl⁻ ion. The density at the point of contact between the ions is practically zero, so that they can be said to be completely separated.

Such results justify a very simple method for calculating the electrostatic interaction between ions, which can be replaced by the charge $+e$ or $-e$ located at their centres. Coulomb's law applies as strictly on an atomic scale as on a macroscopic one and on these simple foundations we are in a position to find *quantitatively* the binding energy of a crystal due to its ionic bonds.

We shall illustrate the method of calculation by taking a purely schematic example: that of a row of positive and negative ions spaced at regular intervals a along a straight line (Fig. 4.29). What we need to find is the work we need to do to stretch out this line of N ions uniformly until they are so far apart that the interactions fall to zero. To do this for two charges q and q' separated by a distance r requires a work equal to $-qq'/4\pi\varepsilon_0 r$. So if we consider first all the pairs of ions containing A_0, that is A_0A_1, A_0A_2, A_0A_3, etc., on one side and the same on the other, the work required to separate all these pairs is

$$W = 2 \times \frac{e^2}{4\pi\varepsilon_0}\left(\frac{1}{a} - \frac{1}{2a} + \frac{1}{3a} - \frac{1}{4a} + \cdots\right)$$

$$= \frac{2e^2}{4\pi\varepsilon_0 a}\left(1 - \frac{1}{2} + \frac{1}{3} - \frac{1}{4} + \cdots\right) \tag{1}$$

The infinite series in parentheses has the value ln 2 so that

$$W = \frac{2e^2 \ln 2}{4\pi\varepsilon_0 a} \tag{2}$$

This is of course the work done in relation to *one* ion A_0, and from it we can deduce the binding energy for N ions. However, there are two remarks to be made. Firstly, if (2) is simply multiplied by N, each pair of ions would be counted twice, once as a partner of A_0 and once with A_0 as a partner. The multiplying factor is thus $N/2$. Secondly, the calculation is clearly not exact because (2) assumes that the series in (1) is infinite, whereas the number of ions is finite. However, only a limited number of terms in (1) will be significant, say about 200, and in that case the sum in (2) is sufficiently accurate for a row of more than 400 ions. Given that N is always large (about one million for a row of 1 μm length), we can legitimately claim that the binding energy of the whole row is $(Ne^2 \ln 2)/4\pi\varepsilon_0 a$ or $(e^2 \ln 2)/4\pi\varepsilon_0 a$ per ion.

Let us now pass to a real case. Consider the structure of a KCl crystal containing K⁺ and Cl⁻ ions arranged as shown in Fig. 4.17 for NaCl with K⁺ replacing Na⁺. The binding energy will be equal

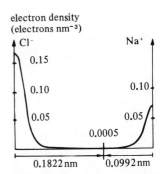

electron density (electrons nm⁻³)

Fig. 4.28 An electron density map for NaCl along the line of centres of the two ions. Quantum mechanics specifies the probability of finding the electrons around a nucleus or, what is the same thing, the electron density at any point (expressed as the mean number of electrons per nm³). X-ray diffraction allows us to *measure* this density. The minimum value in the figure is only one-thousandth of the maximum value at the centre of the ions. In the sphere of radius 0.1822 nm centred on Cl⁻ there are 18.25 electrons, and in a similar sphere of radius 0.0992 nm round Na⁺ there are 9.56 electrons. This justifies our picture of completely separate ions: Cl⁻ with 18 electrons and Na⁺ with 10. (G. and J.-P. Vidal, Faculty of Science, Montpellier.)

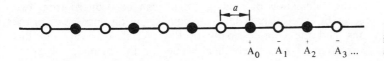
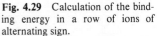

Fig. 4.29 Calculation of the binding energy in a row of ions of alternating sign.

to the work needed to disperse the ions of the crystal. Just as in the linear row, this will be obtained by summing the work needed to separate each pair of ions until they are an infinite distance apart. The calculation is much more complicated here because the lengths of all the distances between pairs must be listed, starting with the shortest and moving outwards. Thus, if a K^+ ion is our reference, there are 6 Cl^- ions at $a/2$, 12 K^+ ions at $\sqrt{2}a/2$, 8 Cl^- ions at $\sqrt{3}a/2$, etc. To dissociate all the pairs in which the chosen K^+ is one of the two requires a total work

$$w = \frac{e^2}{4\pi\varepsilon_0 \frac{1}{2}a} \left(\frac{6}{1} - \frac{12}{\sqrt{2}} + \frac{8}{\sqrt{3}} - \frac{6}{\sqrt{4}} + \cdots \right)$$

This summation can be carried out numerically, although it is a lengthy process because the series converges very slowly. It is found that

$$w = \frac{2e^2}{4\pi\varepsilon_0 a} \times 1.747$$

Using the same reasoning as with the linear chain, the multiplying factor for N ions is $N/2$ and thus the binding energy *per ion* is $w/2$. For KCl where $a = 0.628$ nm, this is found to be 0.64×10^{-18} J. A mole of KCl contains 6.02×10^{23} K^+ ions and the same number of Cl^- ions, so that in the end we find that the binding energy is

$$W = 770 \text{ kJ mole}^{-1}$$

This is a result that can be compared with experiment. When sublimation of KCl occurs, the crystal splits up into dispersed ions (just as assumed in our calculation) and these are then converted into atoms by the reactions

$$K^+ + e^- \rightarrow K \qquad Cl^- \rightarrow Cl + e^-$$

for which the energy is known. (KCl vapour is composed mostly of K atoms and Cl atoms with very few KCl molecules, which we neglect.) The heat of sublimation, which can be measured, less the energy needed for the conversion of ions to atoms, should give us an experimental value for the binding energy of the crystal. The value obtained is 685 kJ mole^{-1}. The agreement with our theoretical value of 770 kJ mole^{-1} is not bad, but we have forgotten something in our model which makes it even better.

In addition to the attractive Coulomb force that we have used, there is, as we have seen, a repulsive force as well, and at the equilibrium distance the two are equal. When the ions are being separated, the repulsive force decreases very rapidly, but it does nevertheless contribute some work with the opposite sign to that of the electrostatic force. It thus reduces the binding energy. The exact law for the repulsive force is not known, but the correction it makes to the calculation of binding energy can be deduced from two pieces of experimental data: the equilibrium distance apart of the ions and the compressibility of the crystal. Carrying out this correction gives a theoretical value for the binding energy of

679 kJ mole^{-1}, in excellent agreement with the experimental value of 685 kJ mole^{-1}.

This result is really quite remarkable because an extremely simple theory is able to explain quantitatively the very essence of a crystal's existence: its cohesion and the energy associated with it. However, it is only for ionic crystals that we can go so far and the reason is that the major contribution in them arises from electrostatic forces, for which Coulomb's law is as perfectly valid on an atomic scale as it is macroscopically. This is not true for the laws of classical mechanics which, in atoms, have to be replaced by those of quantum mechanics: unfortunately exact calculations here are often not practically possible. So it is not only in molecular crystals that the binding energy cannot be calculated accurately from theory, but in other types as well that we shall meet shortly.

In the model of an ionic crystal, each ion is linked to quite a number of neighbours by identical bonds. The structure is thus quite different from that of a molecular crystal. There is no reason to associate a K$^+$ ion with one of the neighbouring Cl$^-$ ions rather than any of the five others: *the KCl molecule does not exist in the crystalline state*. In fact, such a molecule does not exist in the liquid state either, because molten KCl is also composed of positive and negative ions—not with long-range order, but alternating over short distances to preserve local neutrality. In the vapour, there are a few KCl molecules but most of the material is dissociated into atoms along with a few more complex groupings.

Covalent crystals

As we saw in Chapter 1, the covalent bond is one of the basic forms of chemical binding *within* a molecule such as H$_2$ or CH$_4$, but it is also encountered in crystals when their 'particles' are neutral atoms. In an elementary way, the bond is represented by a pair of electrons *shared* between two atoms and so is called *homopolar*, in contrast to the *heteropolar* bond between ions of opposite signs characteristic of ionic crystals.

The binding energy of covalent bonds can be calculated, and a typical result is the value quoted in Chapter 1 for the H—H bond in H$_2$: 458 kJ mole^{-1}. This makes it a *strong* bond with an energy comparable to that for ionic binding. However, we again emphasize that whereas the latter can be explained by classical theory, only quantum mechanics can account for the covalent bond.

When an atom has an external shell which is incomplete because it lacks one or more electrons, it can be covalently linked to several neighbours. All neighbours of the same species will be at the same distance, but the essential feature is that the bonds are *directed*—in other words, the angle between interatomic vectors is fixed. For instance, carbon (p. 6) has four electrons in its outer shell (which can accommodate eight before being saturated) and so it can be linked to four neighbours. The valence bonds are directed towards the corners of a regular tetrahedron with the carbon itself at the centre.

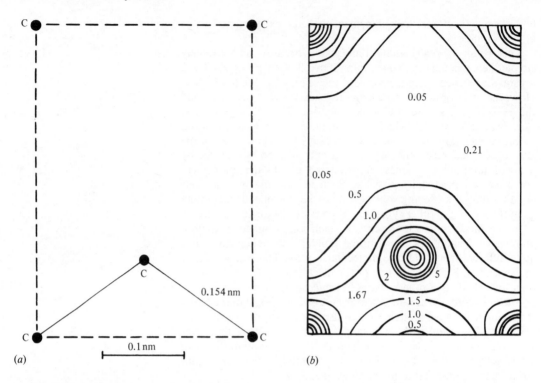

(a)

(b)

Fig. 4.30 Electron density map in diamond. In the lattice plane shown in diagram (a), a carbon atom is linked by covalent bonds to two neighbours at a distance of 0.154 nm. In (b), the experimental electron density is given by contours at heights of 0.5, 1.0, 1.5, etc. The minimum value along the line of the bond is a saddle of height 1.67. Unlike the case of NaCl, the density does not become negligible along the bond between atoms. (S. Göttlicher and E. Wölfel, *Zeit. Elektrochemie.*)

This is the arrangement of carbon atoms in a diamond crystal, where the four neighbours are themselves all carbon atoms. An electron density map, obtained by using X-ray diffraction methods, clearly demonstrates that the carbon atoms are linked by a *bridge* of electronic charge (Fig. 4.30). It has been calculated ·by G. and J.-P. Vidal of Montpellier that a sphere of diameter 0.154 nm (the length of the C—C bond) centred on a carbon nucleus contains only 4.48 electrons instead of the six electrons of the neutral C atom. There is thus a distributed electric charge outside these spheres (which are not close-packed in diamond), but it should be noted that the amount of charge associated with the bond itself between the C atoms is only 0.2 electron. This is a long way from the pair that are shared in the simple model of the covalent bond. In spite of that, the small negative charge between the atoms is the source of the very considerable binding energy of diamond.

The structure of a diamond crystal, also obtained very easily from X-ray diffraction experiments, shows each carbon atom surrounded by four neighbours at the corners of a regular tetrahedron, with each of these surrounded by four neighbours similarly disposed, and so on. The structure can be extended indefinitely by repetition, *all the bonds between neighbouring atoms being covalent.* In a molecule such as CH_4, the linked group of atoms is finite because all the covalent bonds in all the atoms are saturated. In a diamond crystal, no matter how many carbon atoms there are, there is never complete saturation and so, in that case, *the covalent crystal can be regarded as a giant molecule of indefinite extent.*

An arsenic atom has five electrons in its outer shell and, since it could accommodate eight, there are three vacant places. The atom can thus link itself to three neighbouring arsenic atoms by covalent bonds. These bonds are found to have an angle of 92.8° between them and the three neighbours cannot therefore be in the same plane as the original atom. In fact, the neighbours form the base of a tetrahedron with the original atom at the apex. The structure, repeated indefinitely in this way as shown in Fig. 4.31, forms a sort of embossed pattern extending indefinitely in two dimensions, forming a sheet of thickness 0.137 nm. Crystalline arsenic has a structure composed of layers like this stacked on top of each other. In this case, the giant molecule containing only covalent bonds is virtually *two*-dimensional, apart from the up-and-down displacement between nearest neighbours. These 'molecules' are linked to each other by Van der Waals forces between layers just as in molecular crystals.

An atom with six electrons in its outer shell, such as selenium, can be linked with $8 - 6 = 2$ neighbours. The bonds make an angle of 103.1° with each other and the giant 'molecule' here is a helical chain as shown in Fig. 4.32. The helices are packed parallel to each other in selenium crystals with Van der Waals bonds between them.

Finally, an atom with seven electrons, such as chlorine, can only link up with *one* neighbour to form the molecule Cl—Cl. This is the same situation as in hydrogen whose outer shell can only accommodate two electrons—like chlorine, it lacks only one to complete it. So chlorine in the solid state forms molecular crystals like H_2—the 'particles' are covalently-bonded Cl_2 molecules bound to each other by Van der Waals forces (Fig. 4.33).

It is thus only those atoms with four valence electrons that can form crystals containing *nothing but* covalent bonds. With any others, crystals contain both covalent and Van der Waals bonds, the latter being by far the weaker.

Metallic crystals

Isolated atoms of metallic elements are characterized by having outer shells containing only one, two, or more rarely three, electrons. These are the ones that are most weakly bound to the nucleus and that can therefore be easily removed to leave a positive ion.

In *normal* metals, the subshell inside the one containing valence electrons is always complete. The ions of the alkali metals, of the alkaline earths, as well as that of Al (Al^{3+}), have the electronic configuration of one of the inert gases. For example, Na^+, Mg^{2+}, Al^{3+} have the same configuration as neon (see p. 10). In addition, Cu^+, Zn^{2+}, Ag^+, etc., also have complete subshells but here there is no inert gas with the same configuration.

In the so-called *transition* metals, there is an incomplete subshell just inside the valence electrons. Thus, in the series Sc, Ti, V, Cr, Mn, Fe, Co, Ni the subshell is progressively filled and becomes complete with ten electrons in copper.

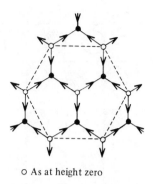

o As at height zero

● As at height 0.137 nm

Fig. 4.31 A wrinkled plane of atoms in a crystal of arsenic: the bonds rise a distance of 0.137 nm perpendicular to the plane in the direction of the arrows. The real angle between bonds is 92.8° and not 120° as it would be if the plane were flat.

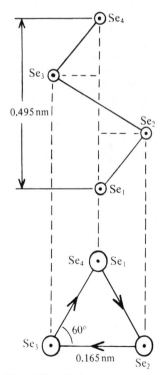

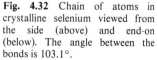

Fig. 4.32 Chain of atoms in crystalline selenium viewed from the side (above) and end-on (below). The angle between the bonds is 103.1°.

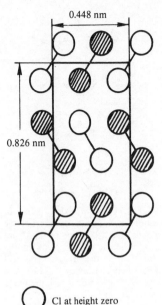

0.448 nm

0.826 nm

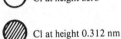

○ Cl at height zero

▨ Cl at height 0.312 nm

Fig. 4.33 The structure of solid Cl_2. One layer of Cl_2 molecules (open circles) packs with inclinations which alternate as shown. The next layer (filled circles) is deposited in the interstices of the first. The distance between two Cl atoms in the molecule is 0.202 nm (compare 0.199 nm in the gaseous form, Table 1.4) but is at least 0.334 nm between different molecules.

What is the origin of the metallic bond? It cannot be ionic because all the atoms are identical and there is no reason why they should form ions of opposite signs. Neither can it be covalent: in covalent compounds an atom is linked to $8 - N$ neighbours, where N is the number of valence electrons. However, since the atom must form a bond using one of the N electrons for every neighbour, it follows that $N \geqslant 8 - N$. So N must be at least equal to four and this is not the case for metals.

The valence electrons are thus too few in number to link atoms to their neighbours by bonds between pairs, and so they become detached from the individual atoms and the whole collection forms a 'sea'. It is this which ensures the cohesion of the metallic ions in the crystal. The detachment of the valence electrons means that the original atoms of the metal are divided into:

1. *Positive ions,* which are regularly ordered in a crystalline array, generally a simple one of high symmetry. We shall return to this point.

2. *The electrons,* which behave like nearly free electrons enclosed in an empty and tightly-shut box. They are confined to the interior of the metal because, if one managed to escape from the surface, it would leave behind a positive charge which would pull it back. The free electrons undergo rapid movements and they can also be made to drift along under the action of an externally applied electric field. The latter movement is, of course, an electric current so that the free electrons are responsible for the *electric conductivity* of the metal. The energy states available to the electrons are not localized around a single atom and are called 'delocalized states': details of their nature and the distribution of the electrons amongst them can only be calculated using quantum mechanics. Although we cannot deal with the theory here, it can be seen that the energy of the whole electron assembly could be calculated and thus the binding energy of the metal—that is, the work necessary to convert the metal crystal into a dispersed collection of atoms all remote from each other. Contrary to what we have seen with ionic crystals, the calculation here can only be approximate.

In addition, it is well known that the properties of the various metals exhibit wide ranges of values, and different metals accordingly have differences in their atomic models. In alkali metals, the positive ion has a diameter markedly smaller than the distance between the centres of neighbouring atoms in a crystal (see Tables 1.1 and 1.3). The ions are so far apart that they have a relatively small interaction and, since the volume *between* the ions represents 89% of the total, the approximation of considering the electrons as free is very good (Fig. 4.34). In this case, the binding energy of the metal is around 100 kJ mole^{-1}, which is rather weak compared with the typical figure of 500 kJ mole^{-1} in covalent crystals but is large compared with the 10 kJ mole^{-1} characteristic of Van der Waals bonds.

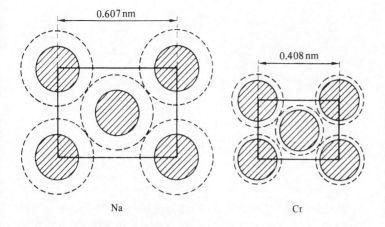

Fig. 4.34 Arrangement of Na and Cr atoms in those planes of the metallic crystal lattice where they are most dense. The broken circles give the sizes of the atoms in the metallic state, while the filled circles correspond to the sizes of Na⁺ and Cr²⁺ in ionic crystals. The volume available to free electrons is much greater in metallic sodium than in chromium.

In transition metals, such as chromium and tungsten, the ions are almost in contact. This gives rise to interactions between the outer incomplete shells of the ions, a little like valence bonds, which make a contribution to the binding energy additional to that from the valence electrons. Thus the binding energy of chromium is about 400 kJ mole⁻¹, while that of tungsten is more than 800 kJ mole⁻¹, one of the largest values observed in any crystal.

Comments on interatomic bonds in crystals

Comparison with intramolecular bonds (i.e. those *within* molecules)
Some of the binding mechanisms between atoms in a crystal that we have discussed also exist within a molecule—in particular, the covalent bond, which is the fundamental chemical bond. Thus, two carbon atoms in a diamond crystal are bound in the same way as in a molecule of a paraffin: even the distances apart are extremely close (0.1541 nm and 0.1544 nm). Similarly, there exist electrostatic forces of the ionic type between atoms in the interior of molecules due to a transfer of charge between them.

However, the pure metallic bond does not exist within a molecule. It is true that electrons are sometimes 'delocalized' to a certain extent: in the quantum mechanical picture, they are shared between several atoms and contribute to their common bonding. In a metal, the valence electrons are *perfectly* delocalized and thus shared by the whole crystal. Through that, they exercise their cohesive action.

Mixed types of bond
In order to simplify the explanations, we have made very sharp distinctions between the various types of bond. This is the moment when we should indicate that reality—as always—is more complicated. To provide adequate representations of some particular cases, we have to make use of *intermediate models* in two categories:

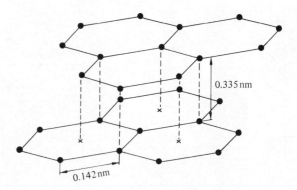

Fig. 4.35 The structure of graphite. Within the planes parallel to the base, the carbon atoms are linked covalently with bonds of length 0.142 nm. The shortest distance between atoms in *different* planes is 0.335 nm.

1. Firstly, *different types of bond can coexist*. We have already indicated that an atom can be linked covalently to certain neighbours (*within* a chain or a sheet) and by a Van der Waals force to others (*between* chains or *between* sheets).

Another example is graphite, whose crystal consists of planes with a hexagonal structure stacked one above the other (Fig. 4.35). Within one plane, each carbon is linked to three neighbours by covalent bonds, using up three of its external electrons. The fourth is delocalized and contributes to a metallic type of binding between the planes. Figure 1.6 provided yet another example: a complex ion formed from several atoms linked covalently but bound to other such complexes purely ionically.

2. Secondly, covalent bonds between two neutral atoms and ionic bonds between ions of opposite sign are only extreme cases of a *whole series of bonds which are partially ionic and partially covalent*. This corresponds to an incomplete transfer of an electron from one atom to another, a concept that would make no sense in classical physics but is possible in quantum mechanics. We speak of the 'degree of ionicity' of the bond as a measure of the amount of transfer, but the precise evaluation of this quantity is difficult. The phenomenon occurs as much between the atoms in a molecule as between those in a crystal.

Can a crystal structure be predicted or explained?

In this chapter, we first described the geometry of structural models of crystals and indicated how they could be determined experimentally. Following that, we examined the interatomic forces that ensure the cohesion of the crystal. The nature of these forces determines the type of atomic arrangement that occurs in a given volume, so that one question immediately faces us: can the structure of a crystal be theoretically predicted from the properties of the constituent atoms and thus from the forces that bind them together?

Thermodynamics indicates the way to proceed. Firstly, if a given substance is at a very low temperature (strictly it should be at absolute zero), *the stable crystal structure is that with the largest binding energy.* So if we know how to calculate the binding energy for all *possible* structures of the substance, the one actually observed is that with the greatest value. Unfortunately, the calculations are not very precise and the differences between values for the various forms are of the same order as the estimated error. The structure with the maximum binding energy cannot therefore be determined with certainty.

To discuss the stability of a crystal at any temperature, we bring in a general rule from thermodynamics: the most stable state of a system is that with minimum free energy $F = U - TS$. Here, U is the internal energy of the system and S its entropy. If the zero for all energies is taken as the state where atoms are completely dispersed, U is none other than the binding energy with a change of sign, $U = -W$. This confirms our argument in the last paragraph, since at $T = 0$, the minimum of F is clearly the maximum of W. The entropy S is a measure of the disorder of the system and in a perfectly ordered crystal at absolute zero it depends only on the thermal vibrations. Now consider two possible forms A and B of the same crystalline substance. At $T = 0$ it is A which is the stable form, say, because $U(A) < U(B)$ as in Fig. 4.36. When the temperature rises, it is the free energies that must be compared. The internal energies U vary little with temperature, so if the entropy of one form, B, is greater than that of A, the curves of F can cut each other. In that case, above a certain temperature T_c the stable form will be B. This argument accounts for phase changes that are observed in numerous crystals, for example α-Fe $\rightleftharpoons$ γ-Fe at 910 °C (see p. 95). However, the calculations of entropy, just as with binding energy, are too imprecise for the transition temperature to be determined accurately.

In the phase equilibrium diagram of Fig. 2.6, the ordered state was assumed to consist of a single crystalline phase. In some substances, however, the situation can be much more complicated. Several phases with different crystal structures can exist, each one stable over a definite range of temperatures and pressures and transforming to another phase outside this range. These are called 'allotropic phases'. By submitting ice, for example, to pressures up to 30 kilobars, six crystalline structures are revealed, with regions of stability as indicated in Fig. 4.37. Conditions at points along the lines of separation between regions allow two phases of ice to coexist. Theory, however, cannot predict the possible structures nor can it determine their range of stability.

In the current state of our knowledge, the crystal structure must very often be regarded as given experimental data. In other words, it is injected into the theory of solids as one of the starting points and used for further development. In spite of that, there are simple cases where the observed structure can be easily predicted or at least explained once it is known. Such successful applications of our theories, even if only partial, are very important factors in increasing our knowledge and understanding of the solid state of matter. We now take a look at some of the most fundamental structures and seek to explain them.

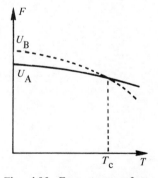

Fig. 4.36 Free energy of two crystalline forms of a substance as a function of temperature. The stable form is the one with the lower free energy so that there is a transition from form A to form B as the temperature rises through T_c.

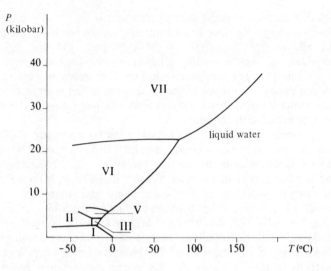

Fig. 4.37 Equilibrium diagram for different allotropic forms of ice. Ordinary ice is phase I, and the others are labelled II, III, V, VI and VII.

Directed and undirected bonds

The first factor that determines a structure is the directed or undirected nature of the interatomic bonds.

1. *Van der Waals*, *ionic* and *metallic* bonds are **undirected**. For a pair of atoms that approach each other, we know that the binding energy increases until a position of equilibrium (or point of contact) is reached. One of the pair could also attract others, which would also tend to come into contact with it, but provided that there is room for them around the first atom, the relative positions of all the neighbours are immaterial. In the whole crystal, the distance between atoms must be the shortest possible: in other words, **maximum compactness** or **closest packing** is the fundamental criterion in determining the structure.

With inert gas and metallic atoms, the hard sphere model is a good approximation. Predicting the structure then comes down to a problem of geometry: to find that regular arrangement of spheres with given radii that has the maximum number per unit volume. We shall show that, when the spheres are of equal radii, there is a simple solution to this problem which accounts very well for observed structures.

The hard sphere model is also satisfactory with ions, but here the packing must not only be as compact as possible but it must also achieve electrical neutrality by a suitable alternation of +ve and −ve charges. In addition, spheres representing different ions have different diameters, so the problem is more complex.

Finally, the molecules in molecular crystals are also packed tightly against each other, but they can have complicated shapes instead of spheres and, in addition, there can be localized interactions (such as electrostatic forces and hydrogen bonds) at certain special positions in the molecule. The criterion of closest packing is no longer sufficient.

2. *Covalent* bonds are **directed**. When an atom is linked covalently to two neighbours, the interatomic bonds must make a fixed angle with each other such as the 109° between two carbon atoms:

$$C \underset{109°}{\overset{C}{\diagup\diagdown}} C$$

This maintenance of the angles between bonds is the principal condition that must be obeyed by the structure: the packing of atoms is no longer the closest possible. For that reason, the most useful model represents not the atoms themselves but the bonds between neighbours, using lines of length and direction that are fixed relative to each other. This was how the diamond structure was presented in Fig. 1.4.

In the great majority of crystals, the simple rules we have quoted are not sufficient to *predict* the structure, but once it has been experimentally determined using X-ray diffraction, it should be possible to check that the rules are obeyed. However, some prediction is possible in a few exceptionally simple cases and since these cover a wide range of materials we now give a few examples of them.

Close-packed structures

The simplest case of all is that of an element, either an inert gas or a metal: all the atoms are identical and can be represented by hard spheres of known diameter (see Table 1.3), while the bonds are perfectly isotropic (undirected). We must first examine the way such spheres can be packed regularly together in as compact a manner as possible.

Theoretical construction of close-packed structures
Our starting point is a straight row of touching spheres of diameter a, forming a one-dimensional lattice having a period also equal to a. A two-dimensional lattice can be formed by a regular addition of similar rows all in contact—but in order to make the whole lattice as compact as possible, the second row of spheres must be fitted into the hollows between the spheres of the first row. Figure 4.38

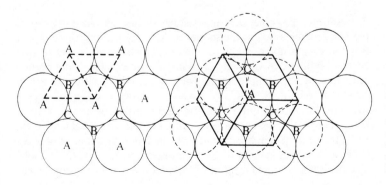

Fig. 4.38 Coplanar spheres of the same diameter in contact form a close-packed hexagonal array (positions A). Each sphere is surrounded by six touching spheres. A second layer could be placed on top of the first so that its spheres lay in the hollows B, or equally well in the hollows C.

shows clearly how this is done and also shows that the lattice obtained is *hexagonal*, like a honeycomb. Each sphere touches six neighbours, which themselves are in contact with others. The centres of three contiguous spheres form an equilateral triangle of side a (AAA in Fig. 4.38).

The hexagonal planes must next be stacked on top of each other with the spheres of the upper one lying in the hollow sites of the lower one in order to make the distance between two consecutive planes as small as possible. The centres of four contiguous spheres are then at the corners of a regular tetrahedron of side a. However, out of the six hollow sites around a sphere A (Fig. 4.38), only three marked B *or* three marked C can be simultaneously occupied. This is because the distance between B and C adjacent to each other is less than a while that between two B sites or two C sites is exactly equal to a.

Let us, then, choose the first two layers to be in positions A and B, where the spheres of the latter are shown as broken circles in Fig. 4.38. We then ask where a third layer can be placed on top of B. The hollow sites provided by B for the third layer could be those directly above A or directly above C—either set is possible, but the structures which result are quite different in the two cases:

1. With the sequence ABA, the third layer is in the same position as the first (strictly it is vertically above it, if Fig. 4.38 is taken as showing the horizontal plane). Figure 4.39 shows the sequence ABA in perspective, and if this continues indefinitely, as it can, *a crystal has been constructed*. The unit cell of such a crystal is an upright prism with hexagonal symmetry, whose base is a rhombus with sides a and angles 120° and 60° and whose height c is twice the distance between successive layers. This latter distance is, as we have seen, the height of a regular tetrahedron of side a, that is, $a\sqrt{(2/3)}$. Hence

$$c/a = 2 \times \sqrt{(2/3)} = 1.632$$

This type of crystal is called **hexagonal close-packed** (h.c.p.) and has a *motif of two atoms*: one at the corner of the unit cell and the other, at a height $c/2$, vertically above the centre of one of the two equilateral triangles forming the rhombic base (which one is unimportant since the intrinsic structure of the whole is the same in the two cases).

2. Another possible stacking sequence, just as close-packed as the first, corresponds to the sequence ABC for the first three layers. For the fourth layer, above C, there are two possible sets of sites corresponding to A or B. By choosing A and carrying on a regular sequence, a stacking pattern with a period of three layers is obtained: ABCABCABC ... . This produces a different crystal from the first one, but it can be described in a similar way as belonging to the hexagonal system. The unit cell would have the same base but its height c would be three times the distance between successive layers instead of two, that is, $c/a = 3 \times \sqrt{(2/3)} = 2.448$. The cell thus contains a motif of three atoms, one at its corner and

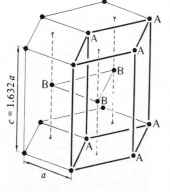

Fig. 4.39   The structure of a hexagonal close-packed crystal, formed by the stacking of hexagonal layers in positions ABA .... .

two others at heights $c/3$ and $2c/3$, vertically above the centres of the two equilateral triangles forming the rhombic base.

However, we saw on p. 59 that the same crystal can always be described in terms of different unit cells. Here we find that the hexagonal crystal can be described with a unit cell of higher symmetry known as a **face-centred cubic** (f.c.c.) cell. It contains a motif of four atoms, one at a corner and three at the centres of the faces of the cube adjacent to the corner. These four atoms form a regular tetrahedron of side a, which we have already identified as the basic component of close-packed structures. Since one side of the tetrahedron forms half the diagonal of the cube, the latter has a side of length $\sqrt{2}a$. Figure 4.40 will help the reader to become convinced that the lattices based on the two unit cells, one cubic and one hexagonal, are identical. However, such figures should not dispense with attempts that should still be made to 'see' the atomic positions in space.

The long diagonal of the cubic cell joining its opposite corners is the same as the height c of the hexagonal unit cell and it is perpendicular to the original hexagonal planes: the opposite corners of the cube at the ends of the diagonal are in the first and fourth of these hexagonal planes. The value of c given by the length of the cube diagonal is $\sqrt{3}$ times the side of the cube which we already know is $\sqrt{2}a$; c is also equal to three times the distance between hexagonal layers, $\sqrt{(2/3)}a$, so that $c = 3\sqrt{(2/3)}a = \sqrt{6}a$ or $2.448a$ as before.

3. There are an infinite number of possible ways of stacking hexagonal planes regularly with periods of repetition that may contain more than three layers. We have so far only looked at sequences whose periods are of two or three layers: ABABABAB ... or ABCABCABC ... . A sequence with four planes repeated indefinitely is ABCBABCBABCB ... .

All the possible arrangements have the same closeness of packing and in all of them each atom is surrounded by twelve neighbours, disposed either as in Fig. 4.41(a) or as in Fig. 4.41(b).

Structures observed in metals and inert gases
How do the structures actually observed compare with the theoretical predictions we have made?

Firstly, we find that all the inert gases, except helium, and some twenty or so metals, including Cu and Ag, crystallize in the face-centred cubic system, *one* of the simplest structures deduced solely from an examination of regular close-packing of spheres. The diameter of the spheres representing the atoms is easily obtained from the experimental value of the unit cell parameter, and the model is thus given a scale.

Secondly, we find that a roughly equal number of metals, including for example Co and Zn, as well as the inert gas helium, belong to the hexagonal close-packed system, which is the *other* simple structure predicted theoretically.

Finally, there are another twenty metals or so, including the alkali metals and α-iron, which do *not* have a close-packed structure but

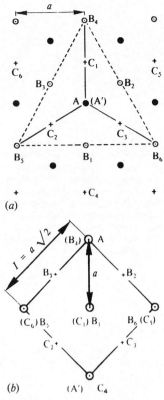

Fig. 4.40 The structure of a face-centred cubic crystal. (a) Stacking of successive hexagonal layers: A (height zero); B (height $c/3 = 0.816a$); C (height $2c/3 = 1.632a$); A' (height $c = 2.448a$). (b) By tilting the tetrahedron $AB_4B_5B_6$ so that AB_4 becomes vertical, the two other sides AB_5 and AB_6 become horizontal. The tetrahedron has a side AB_5 of $I = \sqrt{2}a$. The centres of the spheres are located at the corners and centres of the faces of the cube with sides AB_4, AB_5, AB_6. Parentheses indicate atoms at a height I.

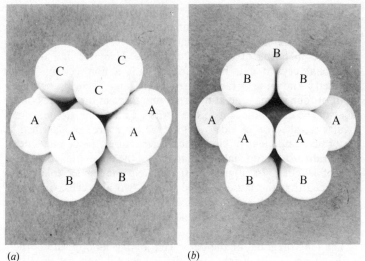

Fig. 4.41 A hexagonal layer of seven balls labelled A (of which four can be seen) with triangles of three balls placed on both sides of it: (a) in non-corresponding sites (balls B and C)—the face-centred cubic arrangement (f.c.c.)—and (b) in corresponding sites (balls B and B)—the hexagonal close-packed arrangement (h.c.p.).

(a)   (b)

each of whose lattice has a **body-centred cubic** unit cell (b.c.c.) as shown in Fig. 4.42. In this, each atom is at the centre of a cube formed by its eight neighbours situated at the corners. If the atoms are represented by touching spheres of diameter a, the side of the cubic cell has a length $l' = 2a/\sqrt{3} = 1.155a$.

In order to compare the body-centred cubic structure with the face-centred cubic one, let us look at the latter in yet another way (remember that we have already seen that it can be described in terms of either a hexagonal or a cubic cell). In Fig. 4.43, the spheres M_1, M_2, M_3, M_4 and M_5 represent those in the base plane of the cubic structure with the centres of M_1, M_2, M_3 and M_4 forming the base of the cell of side $I = \sqrt{2}a$. The spheres N_1, N_2, N_3 and N_4 are at the centres of the vertical faces of the cubic cell and form the next layer above the Ms (note that the close-packed *hexagonal* layers are formed by diagonal sets like M_2, N_1 and N_2). Now there is another unit cell that also represents this structure—a rectangular parallelepiped with a square base of side a ($M_1M_5M_4M_6$ outlined by a broken line) and a height equal to $I = \sqrt{2}a$. We now squash this cell so as to decrease the height I and expand the square base, at the same time keeping the central sphere in contact with the eight others. Eventually, when the height has

Fig. 4.42 The structure of a body-centred cubic crystal. (a) Looking perpendicular to the basal plane: one sphere at the centre of the cube touches four spheres at the corners of the square base (and of course four more on its near-side not shown). (b) A section of the cube taken in the vertical plane through the diagonal AC of (a). A'C' would belong to a nearer set of spheres in (a). The side of the cube can now be seen to be $l' = 2a/\sqrt{3}$.

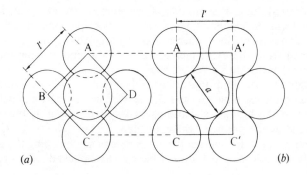

(a)   (b)

become equal in length to the side of the square base, we arrive at the body-centred cubic cell. The b.c.c. structure is only *nearly* close-packed. The fraction of the total volume occupied by the spheres is 0.68, whereas the maximum possible proportion is 0.74 as in the h.c.p. and f.c.c. structures.

We see then that the models involving close-packed or nearly close-packed spheres describe the structures of all the inert gases in their solid form and those of a large proportion of metals—some two-thirds of them. This represents a great success for such a simple geometrical model. However, it does not explain either the *reason* for the stability of one of the close-packed forms rather than another or the frequent existence of the b.c.c. structure which is a little less closely-packed. In practice, the differences between the binding energies of the three structures are relatively small, so small indeed that a number of metals 'hesitate' between two of the structures. Thus, cobalt is h.c.p. at low temperatures but at 400 °C it transforms to f.c.c.

Carborundum, SiC, is formed by the stacking of identical hexagonal close-packed layers—but, in addition to the normal f.c.c. and h.c.p. structures, specimens occur with sequences of various sorts which are much more complex. Repeating motifs are known with sequences of up to one-hundred layers and even more. This is a curious phenomenon known as *polytypism* for which there is no simple explanation.

Iron, body-centred cubic at ordinary temperatures (α-iron), becomes face-centred cubic at 900 °C (γ-iron). This is an example of a phase transition such as we mentioned on p. 87 and its existence determines all the technology of the iron and steel industry: in other words, it has had a considerable importance in the development of our civilization. However, as we have said before, although knowledge of such transitions is essential to the proper use of one of the most common metals, we cannot explain by simple arguments why such particular events occur in the way they do.

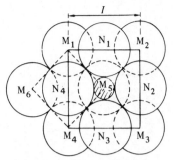

Fig. 4.43 Face-centred cubic crystal with parameter $I = \sqrt{2}a$. The layer of spheres M is the base plane of the f.c.c. cell as seen in Fig. 4.40(b), while the layer N is at a height $I/2$ and at the centres of the side faces. There is a further layer M at height I, and so on. The crystal can also be described by a rectangular parallelepiped with a square base of side a, $M_1M_5M_4M_6$, and with height I. By squashing the height I of this cell, so that the central sphere remains in contact with its eight neighbours, it can be reduced to a cube which is body-centred and has a distance between its M layers of $2a/\sqrt{3}$ or $1.155a$.

In the original f.c.c. lattice, the centres of N_1, N_2, N_3 and N_4, plus those of the M_5s both above and below the Ns, form a regular octahedron of side a. In the central interstice, a small sphere (shown shaded) can be inserted so that it touches the six spheres round it and it will have a diameter of $b = 0.41a$ (that is, $a(\sqrt{2} - 1)$).

Metallic alloys

We turn from the structures of elements to those of crystals containing more than one type of atom, looking first at metals only. In practice, the metals we use every day are mostly not pure elements at all, but are *alloys*. These are obtained simply by melting together the constituents in the desired proportions to produce a homogeneous liquid mixture which solidifies on cooling. Over thousands of years, workers in metals have succeeded by purely empirical methods in making alloys with diverse and remarkable properties—materials that have marked the development of civilization by names such as, for example, *the Bronze Age*. Nowadays, the art of the metallurgist depends on a knowledge of the atomic structure of alloys, which can be very complicated.

Very often, an alloy is a mixture of several 'phases', that is, of different types of crystal with sizes that vary greatly from a micrometre to several centimetres, each being characterized by a definite composition and structure. In some cases, the composition corresponds to a formula similar to that of a chemical compound—for example, $MgZn_2$, Mg_2Si. In general, however, for a given crystal structure, the composition of a phase can vary continuously between quite broad limits. Thus, there is a f.c.c. phase Cu_xAu_{1-x} in which x takes all the values from 0 to 1; a b.c.c. phase Cu_xZn_{1-x} (for $0.46 < x < 0.51$); a cubic phase $Mg_{1-x}Cu_x$ (for $0.65 < x < 0.68$). The formulae of such phases give only their chemical compositions and *do not correspond to molecules*. In the body of the crystal, the different species of atom are mixed up without any order in proportions which, as mentioned above, can vary between certain limits; but the positions of all the atoms, whatever their nature, are fixed just as in a crystal of a pure substance. Such a material is called a **solid solution**. The normal sense of the word 'solution' is that of a liquid consisting of a disordered mixture of different molecules. Here the sense is enlarged to include the solid state.

When we examined the considerable diversity and complexity of metallic phases, it was the atomic models which gave us a systematic classification of the elements and an understanding of their structures. Later in the chapter, we shall give examples that show how the models can explain some of the properties of several well-known metals. We must not ignore the fact, however, that alongside such properties which can be deduced from quite general concepts, there are many more that simply have to be introduced as given experimental data—for the moment, we have to accept that they cannot be explained by appealing to our general models of the structure of matter.

Interstitial solid solutions: iron–carbon alloys or steels

When touching spheres are packed in a regular array as closely as possible, there still remain significant holes or interstices between them—significant because their volume is around one-quarter of the total. In a f.c.c. crystal, for instance, we have seen that there are important groups of six spheres whose centres are at the corners of a regular octahedron (Fig. 4.43). Each sphere of diameter a is in contact with four neighbours and at the centre there is an empty space which can accommodate a smaller sphere without upsetting the octahedron *provided that its diameter is less than 0.41a*. Such a space is called an *interstitial site* since it may be filled by another atom. In the whole crystal, there are as many interstices of this type as there are atoms. It is apparent, in fact, from Fig. 4.43, that in considering a plane of spheres M there is an alternation of a sphere and an interstice along each row M_1M_2, M_6M_5, M_4M_3.

If an additional atom in such an interstitial site has a diameter greater than $0.41a$, it will produce a local distortion of the lattice by pushing out the atoms that it is touching. However, provided that the perturbation is only produced in a small fraction of the sites, the lattice can bear it without the destruction of the long-range order corresponding to the average structure.

An iron atom has a diameter of $a = 0.248$ nm, while that of carbon is 0.154 nm, 50% greater than $0.41a$ ($= 0.10$ nm). It is known that in the f.c.c. structure of γ-iron up to one atom of carbon for every ten of iron can be introduced in an interstitial site. It is true that γ-iron is only stable above 910 °C and no doubt this is because, at that temperature, the atoms have a large amplitude of thermal vibration. The disorder due to this will allow the lattice to cope more easily with the local distortions. Below 910 °C, the iron lattice changes and becomes b.c.c. (α-iron) and it is then observed that the carbon cannot remain in interstitial solution in the lattice. This is a little surprising because the b.c.c. lattice is less closely-packed than is the f.c.c. one. In fact, the interstitial sites have a very different shape in the two structures. In the b.c.c. structure, the largest sphere that can be introduced without distortion has a diameter of $0.16a$ (Fig. 4.42), and carbon, with its diameter of $0.63a$ would cause too great a disturbance in it. A given amount of carbon can thus be in equilibrium in γ-iron but in excess in α-iron, and what happens to it is at the very basis of the structures of steels and the variations in their properties produced by heat treatments—hardening by quenching, softening by tempering, and so on.

It is clearly only rather small atoms such as C, N, B and H which can form interstitial solid solutions. The atomic models make it easy to understand why some metals, in spite of their close-packed structures, can absorb quite large quantities of interstitial atoms. This is a method widely used in metallurgy to improve the properties of a metal: for example, the nitriding of the surface to produce case-hardened steel. Unfortunately, there are also cases where an interstitial element is detrimental: hydrogen, for example, makes a metal brittle.

Substitutional solid solutions: copper–zinc alloys or brasses
If a metal B is to be added to a metal A to form an alloy and the two metals have comparable atomic diameters, B cannot occupy interstitial sites in the structure but *it can be substituted for A atoms*. The alloy then has a lattice as that of the pure metal except that the nodes are occupied by either A *or* B atoms playing the same role as played by the single atom in the pure metal. If the structure is to be at all possible, the difference between the atomic diameters of A and B must always be small (less than 15%, according to a not very rigorous empirical rule). The A lattice then tolerates the local distortions produced by the B atoms up to a limited concentration known as the *solubility limit*, beyond which the excess B atoms form a second phase. This is similar to the way in which water dissolves salt until it is saturated, and beyond this any undissolved salt remains in equilibrium with the solution. As a general rule, the solubility limit increases with rise in temperature because the thermal vibrations make local distortions more tolerable. In the alloy Al–Mg, for example, up to 15% Mg can be dissolved in the Al at 400 °C, but only up to 2% can be dissolved at 100 °C. The difference in atomic diameters is

$$\frac{a_{Mg} - a_{Al}}{a_{Al}} = \frac{0.320 - 0.286}{0.286} = 12\%$$

If the metals A and B have the same valency, then their ions have the same charge and so their electrostatic effects are identical and are the same as in the pure metal A. The total number of valence electrons also remains unchanged, so the metallic bond produced by these electrons is not noticeably affected. In cases where the valency of the dissolved metal is *different* from that of the solvent, there are perturbations in the free electron density to be added to those due to the displacement of A ions by B ones and this can further reduce the solubility. In copper, for example, which is monovalent, gold is soluble in all proportions since it is also monovalent. However, for metals with greater valency there is a solubility limit: with Zn^{2+} it is 38%, with Ga^{3+} it is 20% and with Ge^{4+} it is 10%.

A crystal of a metallic solid solution, whether interstitial or substitutional, has *on average* a structure roughly similar to that of the pure metal. However, differences between the solute and solvent atoms do produce local distortions of the lattice and these confer on the alloys certain properties that are different from, and often better than, those of the pure metal. That is why many alloys have been in use for a very long time.

Steel (iron–carbon alloy) is superior to pure iron in its hardness and mechanical strength. In the same way, brass (copper–zinc alloy) is much stronger than copper. Again, the introduction of copper into gold or silver in coins and jewellery is partly, at least, to increase their hardness. In a quite different area, alloys are often used instead of pure metals because of their greater electrical resistivity: for example, 'nichrome' wires in electric fires are alloys of nickel and chromium.

Molecular crystals

Molecules are attracted to each other by Van der Waals forces and tend to form quite compact arrays, but, because they often have complicated and very irregular external shapes, the geometry of their arrangements cannot have the simplicity of close-packed arrays of spheres. There is also another factor involved. Molecules are electrically neutral, so that the charges in them are balanced overall. That does not prevent the possible production of an electric field around them, however, and such a field would act on the electric charges of the neighbouring molecules. The magnitude of this electrostatic interaction depends on the relative positions of the molecules, which must therefore be such that the electrostatic binding energy is a maximum. It is virtually impossible to *predict* the structure of a molecular crystal, but empirical parameters have enabled us to explain the occurrence of certain frequently observed structures.

In numerous organic crystals, for instance, structures of the type shown in Fig. 4.44 are observed. The elongated molecules form planes in which they are arranged in a herring-bone pattern, that is, with alternating inclinations. This is because the binding energy is greater with this arrangement than when they are all parallel. In

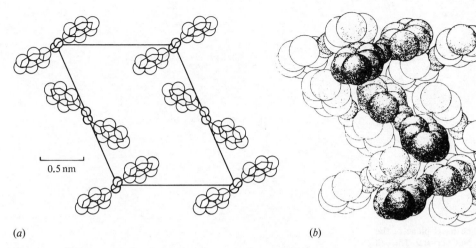

(a)   (b)

addition, successive planes are stacked in a staggered fashion so as to allow the hills and valleys of each plane to fit closely into the adjacent ones and thus to increase the compactness.

Fig. 4.44 Structure of a crystal of azobenzene $(C_6H_5N)_2$: (a) planar arrangement of molecules in one layer and (b) showing the close-packed stacking of molecules. (R. Wyckoff, *Crystal Structures*, Interscience.)

Ionic crystals

The alkali halides
In the alkali halides, such as NaCl, KBr, LiF, etc., the binding is typically ionic and the formulae are the simplest possible, with one ion of each sign. We have already discussed how a regular alternation of +ve and −ve ions ensures the cohesion of the whole arrangement. A particular feature of these cubic structures (Fig. 4.17) is that ions of different sizes have to be packed together. Indeed, Table 1.1 shows that anion (negative ion) diameters are significantly greater than cation (positive ion) diameters to such an extent that it is possible to regard some ionic crystals as formed from the close-packing of the larger anions while the cations are situated in the interstitial sites and provide the electrostatic binding forces.

That is not the case in NaCl because the ratio of diameters a_{Na^+}/a_{Cl^-} is not quite small enough. So there are +ve and −ve ions in contact (Fig. 4.45) and the side of the unit cell, a, is equal to the sum $a_{Cl^-} + a_{Na^+}$. If the Cl^- ions alone were close-packed, the side of the cell would be $\sqrt{2}a_{Cl^-}$ or $1.41a_{Cl^-}$. For Na^+ ions to sit interstitially within the Cl^- lattice, a_{Na^+} would have to be no greater than $0.41a_{Cl^-}$, whereas it is in fact $0.55a_{Cl^-}$. The anions do occupy an approximate f.c.c. arrangement but are a little more widely separated than that: the distance between two neighbouring anions is 0.397 nm (Fig. 4.45) instead of $2a_{Cl^-} = 2 \times 0.181$ nm = 0.362 nm.

Most of the alkali halides have the same structure as NaCl but some, such as CsCl, have a different one which is known to occur when the ratio of the cation and anion diameters is significantly greater than 0.41. The anions would then have to be much further apart than in the close-packed structure and so they adopt another

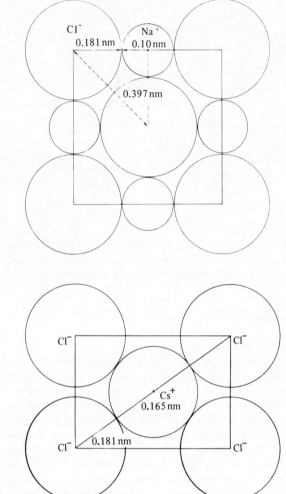

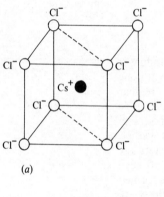

Fig. 4.45 The basal plane of the unit cell in a NaCl crystal. The cell is drawn with relative ionic radii taken from Table 1.1

(a)

(b)

Fig. 4.46 (a) The cubic unit cell of a CsCl crystal. (b) The section of the cell through a diagonal plane of (a) shows the close-packing of the Cs^+ and Cl^- ions.

one altogether: they are situated at the corners of a cube at the *centre* of which lies the cation (Fig. 4.46). Each ion is no longer surrounded by six ions of opposite sign, but by eight, and the alternation of +ve and −ve ions is still preserved even though the structure is very different from that of NaCl.

We are not in a position to make an unambiguous theoretical prediction as to which type of structure, NaCl or CsCl, would be adopted by a given alkali halide. The difference in binding energies of the two structures using the same pair of ions is only about 1% of the total energy. Since this is less than the range of possible errors in calculation, we are unable to tell with certainty which is the greater.

Other ionic crystals
When we pass from the simple cases of the alkali halides to more complicated ionic crystals, we must expect to observe structures

which are not predictable. However, they all obey a simple general rule: ions can be represented by spheres of fixed diameter arranged in such a way as both to minimize the total volume as well as to preserve electric neutrality over regions as small as possible. In other words, the structures must always consist of ions of opposite signs in contact. From an analysis of numerous experimentally determined structures, the following practical rule, known as Pauling's rule, has been derived.

Around a positive ion, there is a layer of negative ions: how many is dependent on the ratio of the radii. These negative ions are in general arranged as symmetrically as possible: four ions in a regular tetrahedron, six in a regular octahedron, etc. Thus the positive charge of the cation is equally shared between the cation–anion bonds, to each one of which is attributed a 'strength' given by the quotient of charge on cation by number of neighbouring anions. In addition, a certain number of bonds from neighbouring cations will terminate on any one anion. Neutrality will be achieved if the negative charge of the anion exactly balances the sum of the positive 'strengths' of the bonds which end on it.

This rule expresses something that is obvious in NaCl, but it is useful as a guide to more complex structures. For example, *beryl* has the formula $Be_3Al_2Si_6O_{18}$. In the structure:

Be^{2+} has 4 neighbouring O^{2-} ions;
$$\text{the Be–O bond has a strength } +\tfrac{2}{4} = +\tfrac{1}{2}$$

Si^{4+} has 4 neighbouring O^{2-} ions;
$$\text{the Si–O bond has a strength } +\tfrac{4}{4} = +1$$

Al^{3+} has 6 neighbouring O^{2-} ions;
$$\text{the Al–O bond has a strength } +\tfrac{3}{6} = +\tfrac{1}{2}$$

In the structure of beryl, the O^{2-} ion has as near neighbours one Si^{4+}, one Al^{3+} and one Be^{2+}. The sum of the strengths of the bonds ending on the O^{2-} ion is $1 + \tfrac{1}{2} + \tfrac{1}{2} = +2$, which exactly balances the -2 of the oxygen. This formal rule is equivalent to minimizing the potential energy associated with the electrostatic field of the ionic charges.

The structures of silicates

When directed bonds occur in a structure, the concept of close-packing of spheres can no longer be used. The essential features are now the interatomic vectors because they determine the *directions* of the valence bonds. Thus, the structure of diamond, the simplest and most typical of all the purely covalent crystals, is explained entirely by the directions of the four valence bonds of carbon. It is interesting to note that, unlike those we have looked at so far, this structure is *not* close-packed but is very open. The relative density of diamond is 3.52, whereas a close-packed crystal made up of atoms with a diameter equal to the C–C distance (0.154 nm) would have a relative density of 7.5.

When crystals contain a large number of different atomic species, or contain bonds with mixed character (covalent–ionic) or even several types of bond, the problem to be faced is no longer the

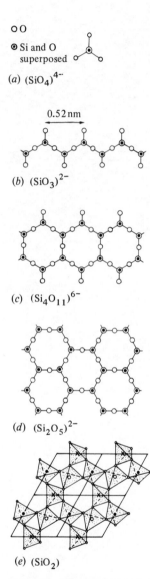

(a) $(SiO_4)^{4-}$

0.52 nm

(b) $(SiO_3)^{2-}$

(c) $(Si_4O_{11})^{6-}$

(d) $(Si_2O_5)^{2-}$

(e) (SiO_2)

Fig. 4.47 The structures of silicates: (a) the fundamental SiO_4 tetrahedron, (b) linear chain of tetrahedra, (c) band formed by two linked linear chains, (d) plane of linked tetrahedra and (e) linked tetrahedra in three dimensions each with four neighbouring tetrahedra—the structure of quartz SiO_2.

prediction of the structure: that is completely beyond the capacity of current theories. More modestly, what is now attempted is the interpretation of experimental results as a means of classifying crystals and understanding their properties.

As an example, we shall discuss the structures of silicates. These form a numerous group, many of which are natural minerals, composed of silicon, oxygen and metallic elements. Their chemical formulae can be very complicated and, to make matters even more difficult, the same species of mineral often has a composition that varies from one specimen to another. For a long time, chemists tried in vain to find a rational classification of silicates by regarding them as salts of different silicic acids, but in the end such a classification has been brought about rather by a comparative study of crystalline structures.

In *all* silicates, the constant elementary feature is the existence of *tetrahedra formed from a silicon ion* Si^{4+} *surrounded by four* O^{2-} *ions, the whole group thus forming a negative ion* $(SiO_4)^{4-}$ (Fig. 4.47(a)). The oxygen is linked to the silicon by a bond whose character is somewhat covalent, but there is another bond at its disposal which can link up with a further, outside, silicon. In this way, two SiO_4 tetrahedra can be linked at one corner by a common oxygen ion. SiO_4 tetrahedra do not link up face to face, and in fact only rarely edge to edge, because these configurations bring the Si^{4+} ions too close for the arrangement to be stable.

The ratio of the number of Si atoms to the number of O atoms varies between 1:4 and 1:2 according to the type of crystal, and this in turn depends on the arrangement of the SiO_4 tetrahedra. These arrangements can be classified as follows.

1. If the tetrahedron is isolated, the silicate consists of two ions: one is $(SiO_4)^{4-}$ and the other is a metal M^{4+}. Example: $ZrSiO_4$ (zircon).

2. Tetrahedra can be associated in small groups: for example, two joined by a corner to form the ion $(Si_2O_7)^{6-}$ or three joined by corners in a triangle to form $(Si_3O_9)^{6-}$, and so on. These complex ions are usually associated with metallic ions that ensure the cohesion and neutrality of the whole crystal. Example: $(Ca^{2+})_2 Mg^{2+} (Si_2O_7)^{6-}$ (melilite).

3. The SiO_4 tetrahedra can be associated in infinite groups, which may be essentially one-, two-, or three-dimensional.

(a) *One dimension* An infinite *chain* of SiO_4 tetrahedra is linked by corners along a straight line such that they point alternately to one side and then the other (Fig. 4.47(b)). The motif has SiO_3 as its formula and it carries a charge -2. This one-dimensional lattice, together with positive ions carrying an equivalent charge, are the elements of a three-dimensional network forming the complete silicate crystal. Example: $CaMg(SiO_3)_2$ (diopside).

Two of these chains, if parallel, can be linked so as to form an infinite *band* whose motif is $(Si_4O_{11})^{6-}$ (Fig. 4.47(c)). *Asbestos* is a silicate of this type and the bands give it its fibrous character.

(*b*) *Two dimensions*   The SiO_4 tetrahedra can be associated by three of their corners to three neighbouring ones in such a way that all their bases are in the same plane. In this way, an infinite layer (a two-dimensional lattice) is formed whose motif has the formula $(Si_2O_5)^{2-}$ (Fig. 4.47(d)). In the complete crystal, these negatively-charged sheets alternate with layers of positively-charged cations. *Mica* possesses this type of structure, which explains its easy cleavage and lamellar texture. In some substances, foreign molecules such as those of water can penetrate between the silicate sheets so that macroscopically the crystal is seen to absorb liquid and swell up. *Kaolin* and other *clays* consist of powders of microscopic crystals having this type of structure: this is the origin of the plastic properties of clay–water mixtures and so is the basis of ceramic techniques that are many thousands of years old.

(*c*) *Three dimensions*   The SiO_4 tetrahedra can be linked together by four corners in a regular three-dimensional array forming a complete crystal. The motif is then SiO_2 and is neutral. This is the arrangement in the structure of a crystallized silica such as *quartz* (Fig. 4.47(e)). It can be seen that the structure contains large interstices so that it is very far from being close-packed. Its rigidity stems from the strong interconnecting Si—O bonds which behave like braces in a mechanical structure in resisting all deformation. (It should be mentioned in passing that in many minerals Al^{3+} can substitute for some of the Si^{4+} ions in the silicate framework, producing what are called alumino-silicates. Of course, the framework of Fig. 4.47(e) is then not neutral, but its negative charge is balanced by small cations such as Na^+, K^+ and Ca^{2+} that can fit into the large interstices. The *felspars* are an example of this type of structure.)

The analysis of the silicates, largely due to W. L. Bragg (1925), was one of the first examples of the help that crystallography has given to the chemistry of solids. It has accounted very easily for the relation between structure and the proportion of Si to O atoms. It has also brought to light the existence of *infinite groups of atoms* which play the role of 'acid radicals' in the presence of metals: it is now clear that these infinite groups cannot be isolated by chemical reactions without breaking up and this has removed a great difficulty that was present in the earlier purely chemical study of silicates.

Some relations between the properties of crystals and their structures

The problems embraced by this title represent the crowning achievement of our treatment of structures of crystals, which include a large proportion of the solids we use and are familiar with. While the knowledge of an atomic structure in itself may satisfy our curiosity, it is only of real interest if we know how to make some use of it. Of course, we have to realize, and accept, the fact that we are far from being able to deduce *all* the properties of a crystalline solid by logical reasoning from its structure combined with the general principles of physics. We can, however, arrive at an understanding of the properties, albeit of only some, rather than being content with just measuring and classifying them, and this alone justifies the lengthy research needed to establish structural models.

The detailed development of methods that use structural models to interpret and understand properties of materials in a comprehensive manner is reserved for another book which would follow on logically from this one. In the following sections, we wish to give merely a few brief qualitative insights into some of the most striking successes of arguments based on atomic models of crystalline matter.

The binding energy of a crystal and its melting point

The melting points of different solids under atmospheric pressure range from a few degrees K to several thousands of degrees. A material is normally classified as a solid or a liquid (or a gas) according to the position of its melting point in relation to the temperature of our surroundings—but, however important this classification may be in practice, it clearly has no fundamental significance.

A crystalline solid melts when the amplitude of the thermal vibrations of its atoms becomes large enough for the crystal structure to collapse. Whatever the nature of the substance, the amplitude then reaches a value which is a significant fraction of the interatomic distance, say several hundredths of a nanometre. However, at a given temperature, the stronger the interatomic bonds which limit the vibration, the smaller will be the amplitude. This inevitably leads to the idea that *the melting point of a crystal will be higher for interatomic bonds with greater energy.*

Table 4.1 shows that classifying materials according to the type of bond does indeed correlate directly with the level of the melting point. At the bottom of the range are the Van der Waals and molecular crystals, such as the inert gases and organic compounds, and right at the top are the covalent crystals. (Diamond does not melt under atmospheric pressure but is transformed to graphite above 3000 °C). Ionic crystals also have a high melting point because of their high binding energy. As for metals, their melting points are spread over a large range from mercury (−39 °C), the alkali metals (from 29 °C for Cs to 180 °C for Li), right up to refractory metals such as molybdenum and tungsten which melt at over 2500 °C. In parallel with this large range of melting points, the binding energies of metals also vary within wide limits (p. 84). Only the valence electrons contribute to the purely metallic bonds in the alkali metals, while in transition metals such as Fe and refractory metals such as Mo there are, in addition, quite strong interactions between electrons of the incomplete inner shell which add to the binding energy.

Table 4.1 Melting points and latent heats of fusion of various crystals.

Melting points (K)	Molecular crystals	Metals	Ionic crystals	Covalent crystals	Molar latent heat of fusion (kJ mole^{-1})
				C (<3500)	
		W (3683)...			35
				BN (3270)	
3000					
		Mo (2883)			
		Pt (2043)			19.6
2000				SiO$_2$ (2001)	
		Fe (1808)..			13.8
				Si (1683)	
		Cu (1336)			13.2
		Ag (1233)		Ge (1240)	
			LiF (1143)		
			KCl (1063)		
1000					
		Al (933) ··			10.8
		Pb (600)...............			4.7
		Bi (544)			
	} Organic crystals				
		Na (371)			2.5
		Hg (234)			2.1
	Cl$_2$ (171)				
0	Ar (84)				

The binding energy of a crystal and its hardness

In ordinary language, one solid is said to be harder than another if it requires a greater force to alter its shape. Mineralogists use a scale of numbers from 1 to 10 to give a more precise but still empirical indication of hardness, and we shall be content to use this fairly qualitative idea to illustrate its correlation with the strengths of interatomic bonds.

Molecular crystals, particularly organic ones, are not hard (1 to 3 on the scale), while, in contrast, covalent crystals are very hard: diamond is at the top of the scale with a coefficient of 10, although a synthetic covalent crystal, boron nitride, is in fact even harder.

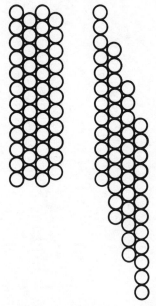

Fig. 4.48 Changes in the external shape of a crystal can take place by the sliding of lattice planes over each other. Notice that the local atomic structure remains unchanged.

Ionic crystals are also in general hard or quite hard. Like covalent crystals, they are scarcely deformed at all under the action of an applied force, whether it be a compression or a tension. When such a force increases beyond a certain limit, however, the crystal breaks. Covalent and ionic crystals are thus *hard* and *brittle* and this is one of the direct consequences of their structures. The strong interatomic bonds at fixed angles make the crystal structure of covalent crystals very rigid. The carbon atoms in diamond, for example, are linked by a network of bonds making fixed angles with each other and so a carbon atom cannot be easily displaced in relation to its neighbours. As a consequence, the whole array cannot be *continuously* deformed: when the applied force becomes too great the bonds, and with them the crystal, break. The structure of ionic crystals is rigid because electrical neutrality must be preserved. Any change in shape is resisted by strong electrostatic forces and so in this case as well *continuous* deformation does not occur.

The behaviour of typical metals such as copper or aluminium is very different. A copper rod can bend without breaking; large rolling mills convert ingots into thin sheets and dies can draw a metal into fine wires. Thus the external shape of the metal can be completely altered, while its volume remains unchanged and it stays in one piece without breaking. The metal is said to be *ductile* and to suffer *plastic deformation*, and this is also a property which is explicable in terms of the structure. The metallic bond is essentially *undirected* so that its only effect is the maintenance of close-packing of atoms, and they can slip over each other without losing contact. If the various planes slip over each other by distances which are whole numbers of atomic steps, the local structure of the metal is the same as before while the external shape has changed (Fig. 4.48).

Electric conductivity and the type of bond

Solids can be crudely classified as either insulators or conductors of electricity. When an insulator is placed in an electric field, no current passes through it because no general drift of electric charge takes place: all the electrons are bound to atoms or molecules by forces which are much greater than those exerted on them by the external electric field. This is the case in molecular, ionic and covalent crystals, in which every electron occupies an atomic or molecular orbital, and it is well known that these substances are indeed electric insulators.

In metals, however, the valence electrons are not attached to any particular atom: they contribute to the metallic bond between the positive ions while moving freely, or nearly freely, through the crystal. An applied electric field exerts a force on the electrons and moves them—in other words, a potential difference causes a current to flow: a metal is a conductor of electricity.

Not only that, but there is also a direct relation between the flow of heat and electricity because both are essentially carried by the free electrons. This explains why a metal is also a good thermal conductor, while thermal insulators are to be found among molecular and covalent crystals.

Optical properties and the type of bond

When a light beam falls on a metal, the free electrons vibrate under the action of the oscillating electric field of the electromagnetic waves in the beam. Each vibrating electron then becomes a small transmitting aerial (antenna) emitting an electromagnetic wave of the same frequency as that of the incident light. The waves from the whole collection of electrons combine to form the reflected wave, to which all the energy of the incident beam is transferred. Nothing penetrates the metal: it is *opaque*. Once again, it is the free electrons that contribute to the metallic bond which are responsible for the reflecting power and opacity of metals or, in other words, for the lustre that makes a metal so instantly recognizable.

This phenomenon does not occur with an insulator because there are no free electrons, so covalent, ionic or molecular crystals can therefore be transparent. This is certainly so with diamond, rock-salt and many organic crystals, although things are in fact rather more complicated than that. Certain electrons, although bound to atoms or molecules, behave like oscillators, but they can only be set into vibration by a light wave if its frequency is close to that of the oscillator (resonance). This means that certain wavelengths of the incident light will be absorbed, while the crystal is transparent to the rest. The transmitted light will have a different composition from that of the incident light and the crystal will appear coloured.

We shall stop our survey at this point, incomplete as it is. It has been kept deliberately superficial because we wanted to demonstrate above all the numerous links that can be made between apparently unconnected facts through the use of our atomic models.

It is the *covalent* bond which is the *common* cause of the *hardness*, the *transparency* and the *absence of melting* in diamond. It is the character of the *metallic bond* that simultaneously explains the *ductility*, the *lustre* and the *electric and thermal conductivity* of metals: a copper rod bends easily without breaking, its surface has a sheen, it conducts electricity and it is cold to touch. So the properties that tell the most non-technical of observers that copper is a metal can be deduced from the way we predict that our atomic model of a metal will behave. However, the largest perspectives are opened up by the fact that theoreticians can predict *quantitatively* certain properties of solids by using the same models.

5 The structure of real crystals

The ideal crystal that we have been describing up to this point is generated by a strict repetition of a unit cell whose shape and contents are well defined. To take account of the perpetual movement of atoms, particularly at higher temperatures, it would be more exact to say that the perfectly periodic arrangements in our models are really exhibited by the fixed centres around which the atoms vibrate.

In fact, real crystals, whether natural or synthetic, do not possess structures as simply defined as this. That should not surprise us since we know that models of reality in physics are always approximate, but in the present case two things are unexpected.

The first is that the majority of 'good' crystals *depart very little from the perfect geometrical arrangements described in the last chapter.* In a pure substance, nearly all the atoms have exactly the immediate environment that they should have, according to the ideal scheme. It has been found possible, for instance, to manufacture single crystals of silicon without any defects at all over a distance of several millimetres, that is, over more than ten million unit cells. Another example is that of the plane faces of a natural crystal, which are found to have the predicted angles between them (e.g. 90° or 120°) to a very high degree of precision. In short, the deviations from the ideal theoretical structure are either very *small* or, if rather larger, very *rare*, with some thousands of perfect unit cells separating the disturbed regions.

The second unexpected point is that, although the defects in a crystal are relatively of little importance in size and number, *their influence on certain physical properties can be considerable.* An analogy will help to understand this major feature in the physics of solids. Suppose we have a chain of 1000 links in which only one is defective. The chain is quite sound over 99.9% of its length, yet its strength is not the average of that of all the links but is much smaller since it depends almost entirely on the single weak link.

To illustrate the effects of crystal imperfections, we give the following few examples (it should not be thought that all the effects are bad, as the word 'defect' would lead us to imagine). Firstly, if one atom in 10 000 were *not* absent from the crystal lattices of metals, brazing (hard soldering) would be very difficult and the heat treatment of alloys in metallurgy would be ineffective. Secondly, if metals did not contain what we shall later call 'dislocations', we could make machines as strong as those produced by the best current techniques but 100 times lighter. Finally, a few well-chosen impurity atoms in a germanium crystal, in a dose of less than one part in a million, has given us the transistor and the extremely small integrated circuits of microprocessors.

These examples illustrate also the source of some major problems encountered in solid state physics. If it is true that some important properties of solids depend on defects, it is essential to know all about them if they are to be controlled and properly used. Yet if they are very small or very rare, it is difficult to 'see' them—in other words, to obtain direct information about their position and their constitution. Moreover, an irregularity, by its very nature, is mathematically difficult to define. We have already mentioned that there are only two simple situations for a theoretician: matter in perfect order or in complete disorder. It is easy to anticipate that a little disorder introduced into a crystal must give rise to very complex, perhaps intractable, calculations. Nevertheless, considerable progress has been made in this field in the last forty years, thanks to new experimental techniques and to more refined theories.

We shall describe in turn the most important types of defect in crystals and indicate for each one the physical properties on which it has an influence.

Vacancies in crystals

A vacancy is a defect resulting from the absence of an atom in a crystal—a node of the lattice is not occupied by the atom that ought to be there (Fig. 5.1).

This type of defect was introduced by theoretical physicists, not because any had been directly observed, but because without their existence it was difficult to account for an experimental fact: that atoms change places in a crystal, i.e. that diffusion takes place. However, let us first satisfy ourselves that a vacancy can exist at all.

A crystal with vacancies has a greater internal energy than a perfect crystal since the atoms are a little less tightly packed than they could be: a certain amount of work must be done to remove an atom from the interior and place it on the surface so increasing the internal energy U. Thus, at absolute zero, and even at temperatures a little above it, vacancies are unstable since equilibrium requires U to be a minimum. At higher temperatures it is no longer the same, because the condition for stability is that the free energy $F = U - TS$ should be a minimum. This means that the *entropy S* is involved and sometimes determines the stable state (see p. 87). Entropy is a measure of disorder in the system and vacancies are of course a disordering feature of crystals. Using statistical mechanics, we can calculate the number of vacancies existing in a crystal in equilibrium at a given temperature. It is found that the concentration is always small: for example, for copper at the maximum temperature just below the melting point, the proportion is 5×10^{-5} and $100°$ lower it is 2×10^{-5}; at room temperature it is completely negligible. Nevertheless, by rapid cooling from a high temperature (quenching) it is possible to preserve the vacancies at ordinary temperatures for up to ten minutes, after which they have

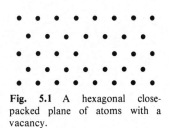

Fig. 5.1 A hexagonal close-packed plane of atoms with a vacancy.

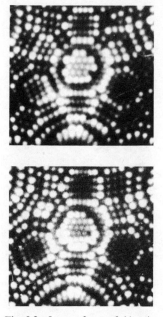

Fig. 5.2 Images from a field emission microscope of the surface of a platinum crystal in the shape of a sharp point. Between taking the two photographs, a certain number of atomic layers have been removed by evaporation. Missing atoms are clearly visible in different places on the two pictures. (J. Gallot, University of Rouen.)

disappeared through the rearrangement of the atoms—the crystal returns to its equilibrium state.

The most direct method of evaluating the proportion of vacancies is by measuring the density but, while it is an absolute method, it is not accurate enough. The most convenient quantity to measure in metals is electric resistivity, a property that is both sufficiently sensitive to the presence of vacancies and easy to measure very accurately.

Direct experimental observation of vacancies would clearly be very desirable as a basis for the theory, but the small size of the defect and its low concentration in crystals make such experiments very difficult. However, we can quote one type of observation that has been made, for a particular case certainly, but of great interest nonetheless. On p. 70 we mentioned an instrument called the field emission microscope. This is capable of giving images of the surface of a very sharp point (diameter 10 nm) with a resolution great enough for individual positions of atoms in the surface planes to be distinguished. In these images (Fig. 5.2) 'holes' corresponding to vacancies can be clearly seen by comparing two photographs and the number that occurs does correspond to the expected concentration.

In spite of the low number of vacancies in a crystal, their presence is vital for the phenomenon of *solid diffusion*. Diffusion in liquids and the very rapid diffusion in gases have already been dealt with (p. 28) and are not unusual. However, in crystals, where the atoms are packed closely together, two of them can only change places by exerting considerable forces on their neighbours. There is thus not much chance of this happening spontaneously, even at high temperatures, yet diffusion *is* observed in solids: at 800 °C, for example, atoms can penetrate some 5 μm in one hour in copper. This is only possible if there are a few vacancies to facilitate the movement. The mechanism can be appreciated from the model illustrated in Fig. 5.3, where the small square pieces can slide sideways or vertically in the plane, provided that there is a vacant space to move into (this is similar to a well-known toy in which the object is to place numbers in a certain order). If there were no empty space, the positions would be completely blocked, but with only one missing, a given piece can cross the whole board in 15 or so moves.

Other point defects

The vacancy is called a *point defect* because it is attached to one point in a crystal, although it does affect a certain number of atoms around it. There are other types of point defect.

Interstitial atoms
Whereas vacancies were formed by missing atoms, these consist of additional atoms. When we talked about solid solutions in the last chapter, we saw that a foreign atom could be sited in the interstices even of a close-packed structure. However, if such an atom were of the *same* type as those in the main structure, the distortion

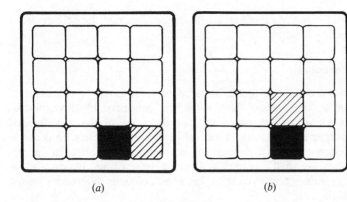
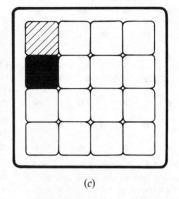

(a)   (b)   (c)

produced would be considerable (Fig. 5.4). This means that the energy needed to get such an atom into an interstitial position is so great that, even at high temperatures, an appreciable number sited in this way would not be produced spontaneously. However, irradiation of a crystal by high energy particles *can* eject atoms out of their normal sites and send them to an interstitial position. In this case, a vacancy and an interstitial are produced simultaneously and this is called a *Frenkel defect* (whereas a vacancy on its own is called a *Schottky defect*). In fact, there are easier ways for an interstitial atom to sit in a lattice than that shown in Fig. 5.4. Very elaborate X-ray diffraction experiments on aluminium bombarded with high energy electrons have recently shown the type of defect illustrated in Fig. 5.5.

Fig. 5.3 A model to illustrate the mechanism of diffusion in a lattice by virtue of a vacancy shown in black. The squares or 'pieces' can slide horizontally or vertically. (a) Initial position: the shaded piece is next to the vacancy. (b) Position of the piece after four moves. (c) Position after twelve more moves. If the piece were on its own, it would take six moves to get from position (a) to position (c).

Impurities

In talking of solid solutions on p. 95, we also said that one type of atom in a crystal could be largely replaced by another if their properties, particularly their diameters, were not too dissimilar. When these conditions are not fulfilled, such a substitution requires a lot of energy. Even then, foreign atoms *can* enter a crystal in small proportions and are then called *impurities*. When crystals are being produced, the starting materials are never completely pure, and there must also be contact with a vessel and with an atmosphere during the production: all these produce a certain degree of contamination and it is extremely difficult to obtain solids with a degree of purity better than, say, a few parts in ten thousand.

Purity is vital in some cases where even very small proportions of foreign material can markedly change an important property. Copper, for example, must be very pure to function properly as an electrical conductor and hence is refined by electrolysis. Again, if aluminium foil is to be satisfactory for household use, the metal must be 99.99% pure: 99% is not good enough. The most striking example is that of the semiconductor industry. This could only develop as it has because it was able to reduce the level of impurities in silicon or germanium to lower than one in ten million, so that chosen foreign atoms could be subsequently introduced at a level of about one part per million, a process called 'doping'. A final example is that of the screens used for detecting radiation and for

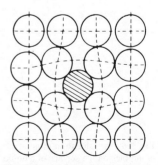

Fig. 5.4 An interstitial atom produces great distortion in the lattice among its immediate neighbours, but the disturbance becomes rapidly smaller at only a small distance away.

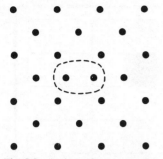

Fig. 5.5 A plane of atoms parallel to the face of the cubic unit cell in aluminium, in which one atom of the perfect lattice is replaced by two. (H. G. Haubold, Julich Laboratory, Germany.)

television receivers, which are fluorescent because they are doped in a suitable fashion.

Dislocations

These are defects that play a major role in the mechanism of metallic deformation. They were first conceived by theoreticians, who could not otherwise account for some of the experimental measurements on the mechanical properties of metals. In order to understand what a dislocation is, and how it can exist in real crystals, we shall first look at a simple demonstration known as the *Bragg bubble raft*, even though it shows obvious differences from a crystal.

Air is blown gently through a tube with a drawn-out end which is placed just under the surface of a dish of water to which some detergent has been added. A string of identical bubbles is produced which tend to cluster together on the surface because of capillary attraction. They form an ordered hexagonal arrangement quite spontaneously, of exactly the sort we described for the most dense planes of atoms in a close-packed crystal (Fig. 4.38). What is more interesting is that *faults* are also spontaneously produced in the arrangement, like that around the point indicated in Fig. 5.6. This defect is certainly not easy to pick out and the raft has to be observed at a grazing angle in a certain direction in order to see the one half-row which stops at the defect, while all the other rows around it remain parallel and complete. Beyond the point where the half-row stops, the void left by the missing bubbles is filled by

Fig. 5.6 A dislocation in an array of bubbles on the surface of a liquid shown on a Bragg bubble raft. To see the dislocation clearly, look at the picture tangentially to the plane of the page in the direction of the arrow. (C. Kittel, *Introduction to Solid State Physics*, Wiley.)

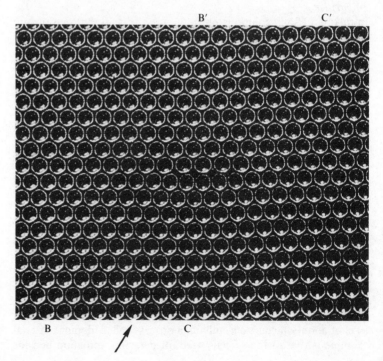

neighbours whose rows are only slightly deformed so that they can come together. As the distance from the defect increases, the displacement of bubbles relative to what their position would be in a perfect lattice becomes smaller. So the fault has effects that are apparently localized in a small volume. Even so, it has effects that are quite clearly evident in distant but apparently normal regions. For instance, if we take the parallel rows BB′ and CC′, there are eight bubbles between B and C, and only seven between B′ and C′.

This demonstration helps us to appreciate what happens in a crystal in which planes of atoms are stacked in parallel layers. One of them may suddenly stop in the middle of the crystal (Fig. 5.7) and the edge of such a plane is called a *dislocation line* (perpendicular to the page in the figure). In the immediate vicinity of this line, the structure of the crystal is distorted, but the 'tube' of matter containing the imperfect arrangement is very narrow: at 0.3 nm from its centre, the displacements of the atoms are of the order of 0.01 nm, and at 5 nm away, the displacements are only of the order of 10^{-4} nm. Outside the narrow tube of distorted structure around the dislocation line, the crystal is very nearly perfect. However, just as in the Bragg bubble raft, the presence of a dislocation can be recognized without leaving good parts of the crystal.

If we had a crystal which was perfect, we could start from any atom A and, by taking a path with equal numbers of steps in opposite directions as shown in the bottom right-hand corner of Fig. 5.8, we could complete a circuit and arrive back at A.

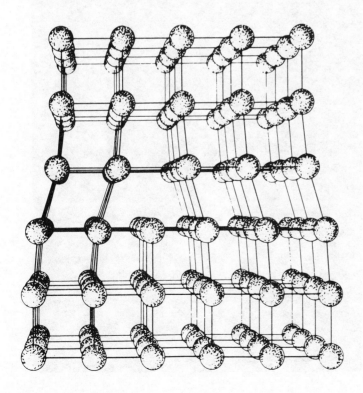

Fig. 5.7 The structure of a dislocation in a crystal with a cubic lattice. The dislocation line is along the edge of the plane that stops half-way up the model and is perpendicular to the plane of the page. (C. Kittel, *Introduction to Solid State Physics*, Wiley.)

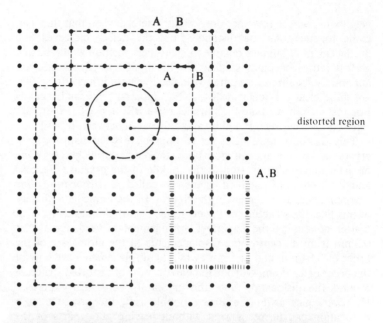

distorted region

Fig. 5.8 The *Burgers vector* of a dislocation. In a perfect lattice, all circuits consisting of equal numbers of steps from left to right and right to left, together with equal numbers up and down, are closed: the finishing point coincides with the start (the circuit ||||||||, B coincides with A). If the path surrounds a dislocation, B does not coincide with A (circuit ————). $\overline{\text{BA}}$ is the Burgers vector which characterizes the dislocation. The region of greatest distortion is inside the circle.

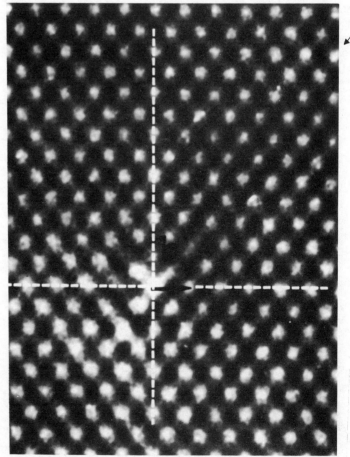

Fig. 5.9 A dislocation in a germanium crystal under a high resolution electron microscope. The distortion at the centre of the dislocation appears more clearly if the picture is viewed tangentially to the page in the direction of the arrow. (A. Bourret, Centre for Nuclear Studies, Grenoble.)

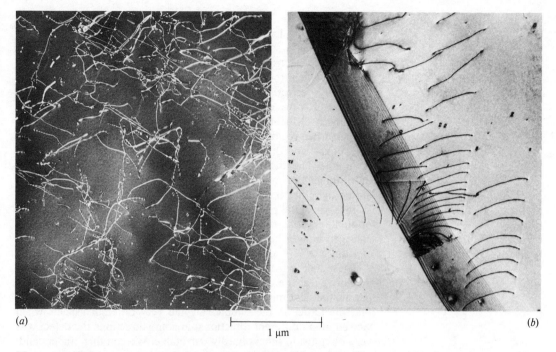

(a)

1 μm

(b)

However, if the same sort of circuit surrounds a dislocation, like one of the upper paths in Fig. 5.8, the finishing point B does not coincide with the starting point A. The vector $\overline{BA}$, called a *Burgers vector*, is independent of the shape of the circuit, provided that it surrounds the dislocation. This vector thus characterizes the defect.

A dislocation is therefore defined geometrically by both the *line* around which the distortion occurs and the *Burgers vector* which is a measure of its total effect on the perfect lattice. The type of defect visible in the Bragg bubble raft of Fig. 5.6 is extended to three dimensions in Fig. 5.7, and it can then be seen that the Burgers vector will be at right-angles to the dislocation line. There are other types of dislocation, in which the Burgers vector is parallel to the line or at an angle to it other than 90°.

Such a model of the defect is a plausible one in that it does not entail intolerable distortions that would raise the energy of the crystal too much. It is, moreover, useful in providing a mechanism to account for the ductility of metals, as we shall see later. However, it is very important that the model should be confirmed by direct observation if at all possible, and this has been done— although long after the work of first theoreticians. We saw (for example in Fig. 4.22) that there are exceptional cases where individual atoms can be resolved by the electron microscope. Such an example is shown in Fig. 5.9, where a dislocation can be seen amid the atoms in a perfect lattice. The interrupted plane of atoms presents a striking analogy to the defect in the Bragg bubble raft.

It is much easier to detect the dislocation *lines* rather than the arrangement of individual atoms in the defect, and this is now within the capabilities of an electron microscope (Fig. 5.10). The lines can also be seen on photographs obtained with X-rays using a

Fig. 5.10 Dislocation lines in pictures from an electron microscope. The specimens are thin sections from: (a) a single crystal of Al–Mg (photographic negative) and (b) two crystallites of stainless steel. The dark band in (b) corresponds to the grain boundary. (B. Jouffrey, Laboratory of Electron Optics, Toulouse.)

technique known as a *Lang topograph* (Fig. 5.11). In both methods, the basic principles are as follows: if a given small volume of the crystal is so oriented that it can diffract incident radiation (X-rays or electrons), the diffracted beam has a much greater intensity when the crystal is slightly distorted than when it is absolutely perfect. Suppose we have a crystal platelet thin enough to allow X-rays or electrons right through it, and the platelet contains a dislocation. If an X-ray photon or an electron crosses the perturbed region along the dislocation line, it is more likely to be diffracted—and thus less likely to be transmitted and form part of the image. The dislocation line thus appears on a photograph like a thin filament of more absorbent material, although obviously the atoms there are the same as elsewhere and have the same density. With X-rays, the pictures obtained can only be the same size as the original crystal, but the electron microscope can yield greatly enlarged images because of its high resolving power (1 nm instead of 1 μm).

These methods locate the dislocation lines for us, and while they do not reveal the distorted atomic structure at the centre, they do permit the determination of the Burgers vector. Figure 5.6 showed that the lattice in the bubble raft had very different aspects when viewed along different directions: in some directions the defect was very obvious, in others hardly detectable. We can then understand, in the same way, why it is that for some directions of the X-rays or electrons, the crystal appears to be perfect in spite of the dislocation and so the line disappears from the picture. By analysing a series of pictures of the same crystal for different orientations, it is possible to deduce the nature of the Burgers vector.

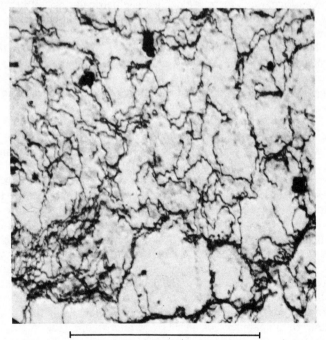

Fig. 5.11 An example of a *Lang topograph*. The use of X-rays provides an image of a LiF crystal platelet that shows dislocation lines in its interior. (A. Authier, Crystallographic Laboratory, University of P. and M. Curie, Paris.)

1 mm

It is thanks to such techniques, which are now standard, that the existence of dislocations can be taken as experimentally established. There are other quite simple methods of detecting and counting them. If the surface of a crystal of silicon is polished and then etched with a suitable reagent, the material is attacked more rapidly in the distorted regions around a defect. Dislocation lines emerging at the surface are the starting points for *etch-pits*, which are easily visible under a good microscope. Figure 5.12 shows silicon crystals from different sources, one with an average of 30 000 dislocations per cm^2 and the other with only 3000. If the number of dislocations is too great, good quality transistors cannot be made from the crystals: this is a typical example of a practical case where it is

(*a*)

⊢——————————————⊣

0.5 mm

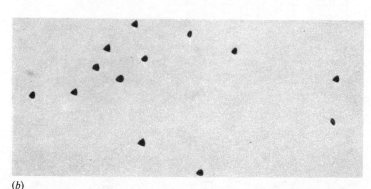

(*b*)

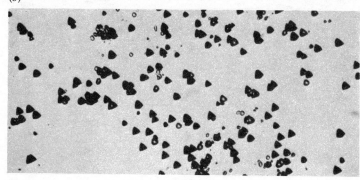

(*c*)

Fig. 5.12 The surface of a silicon crystal seen under an optical microscope after polishing and etching with a suitable reagent. Each etch-pit corresponds to the end of a dislocation line emerging at the surface: (a) crystal without dislocations, (b) 3000 dislocations per cm^2 and (c) 30 000 dislocations per cm^2. (A. Huber, Thomson-CSF Laboratory, Corbeville.)

essential for the semiconductor technologist to detect defects that completely escape the normal crystallographic examination because they are too subtle to show up.

When the crystal is very imperfect, the dislocation lines cluster closely together and can get tangled up in confused skeins. Quite often, however, they adopt a much more regular pattern. For instance, dislocations with the same Burgers vector can be observed to congregate in the same plane and form an ordered array (Fig. 5.13). The tendency of a material to become ordered is so strong that when a crystal contains many regions with few defects, a regular pattern develops. Dislocations collect in a plane separating two parts of the crystal which are each perfect and only slightly

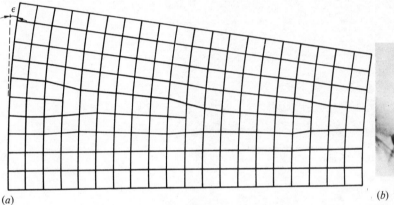

(a)   (b)

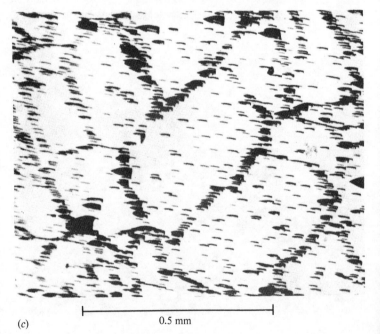

Fig. 5.13 (a) Diagram showing a set of aligned dislocations forming a grain boundary and producing a small difference in orientation ε between two sub-grains. (b) A network of dislocations in the plane of a grain boundary viewed under an electron microscope. (c) Sub-grains on the surface of an aluminium crystal viewed under an optical microscope. Dislocation lines are revealed by etch-pits and can be seen to gather in the grain boundaries. (P. Lacombe, Metallurgical Laboratory, University of Paris-Sud.)

(c)   ⊢————————————————⊣
                                    0.5 mm

differently oriented. This plane is a *sub-boundary* between two sub-grains.

The difference in the orientation of the sub-grains on either side of a sub-boundary is typically only of the order of a few minutes of arc and is even smaller when the dislocations are less dense. The existence of this type of crystal imperfection had been recognized before the concept of the dislocation was introduced. What is considered macroscopically as a single crystal really consists of a *mosaic of sub-grains* separated by boundaries containing dislocations. The scale of the mosaic varies considerably, both as to sub-grain size (from μm to cm) and as to the angle of disorientation (from a few minutes to a degree). Thus, the face of a block of rock-salt appears at first sight to be perfectly plane, as it should be, but closer examination reveals that it is generally covered in facets inclined at small angles to each other. The mosaic structure originates from small local disturbances during the growth of the crystal, and by carefully eliminating these, it has proved possible to produce single crystals of silicon that are absolutely perfect over several centimetres.

The characteristic property of many metals is their *ductility*. Under the action of strong distorting forces, such as those involved in rolling or drawing, a piece of metal changes its external shape while retaining the same atomic structure. The deformation of the metal occurs through a mechanism known as *slip*.

In the last chapter, we saw that a large number of metal crystals were formed from dense layers of atoms stacked in such a way that the atoms of one sit in the hollows of its neighbours in a close-packed array. As Fig. 4.48 shows, the layers can be made to slide over each other and after an integral number of steps will recover the same local atomic arrangement as before. The external shape has changed, but the volume is still the same and the two arrangements are equivalent from the energy point of view because the metallic bond merely encourages the closest possible packing. Nevertheless, between the initial and final equilibrium states, two layers in contact have to pass through an abnormal position (Fig. 5.14), which cannot be achieved without a shearing stress above a certain minimum value to produce the slip. This minimum value can be theoretically evaluated using the atomic model of the crystal. What the calculation shows is that *the minimum shearing stress needed to initiate slip is about 100 times greater than that which is actually observed.* If the crystal is perfect, there is no way, even approximately, of accounting for the mechanical properties of metals, properties that are essential for their practical utilization.

That is where dislocations come into the picture. The slip between two atomic layers is equivalent to the displacement of the dislocation along the plane between them, and this is very easy. *The imperfection thus plays the key role.* It is a complex matter to make this role quite clear and to make quantitative predictions of its effects, but that does not mean that our certainty about the experimental facts is less secure. In fact, the electron microscope shows the movement of dislocations when a specimen is deformed, a movement that can be followed by recording the images on film.

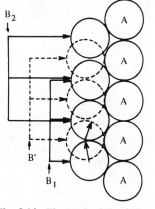

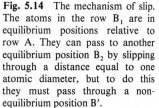

Fig. 5.14 The mechanism of slip. The atoms in the row B_1 are in equilibrium positions relative to row A. They can pass to another equilibrium position B_2 by slipping through a distance equal to one atomic diameter, but to do this they must pass through a non-equilibrium position B'.

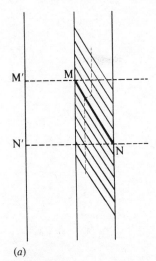

(a)

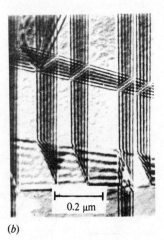

(b)

Fig. 5.15 (a) A stacking fault in a close-packed crystal. The plane of the fault in the crystal slice, MN, produces contrast in the image between M′ and N′ which takes the form of interference fringes. (b) An example of the image produced by stacking faults in the electron microscope. (B. Jouffrey, Laboratory of Electron Optics, Toulouse.)

Stacking faults; twins

If we look at the close-packed structures described in Chapter 4, we find that a pair of consecutive planes taken in isolation always has the same intrinsic structure, whatever the mode of stacking (f.c.c. or h.c.p.). It follows that they also have the same energy, and differences in energy only appear in groups of more than two layers—even then, there is only a relatively small difference in energy between the various possible close-packed configurations. As a result of this, the rules governing the order in which atomic planes occur could be accidentally broken in a crystal grown by the deposition of successive layers.

For example, in a f.c.c. crystal, the sequence should be ABCABCABCABC..., but abnormal sequences will also be found, such as ABCABCBABCABC ... or ABCABABCABC.... Defects of this type are called *stacking faults*. They leave the immediate surroundings of atoms unaffected in all respects, but introduce odd groups of three layers with the h.c.p. sequence into a f.c.c. structure, and vice versa. It is easy to understand from this how such faults will occur very readily when a crystal, such as that of cobalt for example, can exist in either cubic or hexagonal form at the same time, proving how small is the difference in energy between the two structures.

Stacking faults show up in electron micrographs (Fig. 5.15). If a thin crystal platelet, which is otherwise homogeneous and perfect, has a stacking fault along MN, there is contrast in the image between that part joining M′N′ and the rest outside these points. This arises from interference between diffracted beams from layers above MN and from those below it. If the fault plane MN has a large enough area, details of the real structure on an atomic scale can be determined from the micrograph. Not only that, but where instruments of exceptionally high resolution are used, so that individual atoms can be located, cases have been found that conform to the theoretical pattern of stacking faults (Fig. 5.16).

If a f.c.c. crystal is grown by the deposition of successive close-packed layers, the sequence ABCABC... is as likely to occur as the symmetrical one CBACBA.... Should the sequence jump from one to the other in the same crystal, a **twin** is said to be produced, for example by the sequence ABCABCBACBA. The two parts of the crystal have the same structure and are symmetrical with respect to a common plane (C), the twin plane (Fig. 5.17(a)). Once again, the high resolution electron microscope has been used to obtain direct images of a twinned crystal (germanium in Fig. 5.17(b)) agreeing exactly with the theoretical pattern.

Examples of twins occur very frequently among natural crystals, although they can have a much more complicated geometry than the simple type we have been describing. The common characteristic of all types of twinning is that the two crystals of the pair have precisely-defined relative orientations (Fig. 5.17(c)).

Some crystals become twinned under the action of an external force. In this case, the deformation of the solid is not the result of a continuous process as it is in growth, but of a sharp transition to the twinned form. When a rod of tin is bent, the appearance of twins is accompanied by an audible noise (the 'cry' of tin).

Volume defects

The defects described so far have been distortions of an ideal structure localized around a *point*, a *line* or a *plane*. There are also defects which extend throughout a small *volume*, typically with linear dimensions of the order of a few nanometres and quite large enough to involve several hundreds or thousands of atoms. An example occurs in alloys with the so-called *Guinier–Preston* or *GP zones*, whose effects on the properties of the alloys are important and easily linked with their atomic structure.

As we explained in Chapter 4 (p. 95), a solid solution is a crystal with atoms in their proper lattice positions, but with each site being occupied at random by one or other of the various constituent species of atom. Normal methods of observation (optical or electron microscopy or X-ray diffraction) do not distinguish a crystal of solid solution from one of a pure metal since the local distortions due to the dissimilar atoms are not perceptible. Now in certain alloys heat treatment can produce considerable changes in properties without any apparent modification of the crystal structure. Thus, an Al–4% Cu alloy cooled very rapidly (or *quenched*) from 450 °C to room temperature is quite soft, but after remaining for a day at room temperature (during which time the alloy is said to *age*), it spontaneously becomes hard. This effect allows alloys of aluminium to be produced that are of great importance in mechanical construction and are particularly essential in the aeronautics industry where there is a great demand for metals that are both light and strong.

If the properties in such an alloy change, something must change in its atomic structure and that effect has been revealed by more refined methods of observation using electron microscopy and X-rays. The proportion of copper that the Al–Cu alloy can tolerate rises with increase in temperature. Thus, at 450 °C, the 4% Cu is dispersed throughout the aluminium, but, at 20 °C, the Al lattice can only accommodate 0.5% Cu. That is because the higher temperature is associated with greater thermal vibrations which allow more local distortion around the dissolved atoms. When saturated salt solution at 100 °C is cooled, excess salt is deposited in the form of crystals. In the case of the Al–Cu solid solution, the excess Cu atoms *ought* to separate out after quenching in the form of large grains rich in copper. However, that would involve diffusion of the Cu atoms through the solid and this is not possible at low temperatures because the movements of the atoms from one site to another are very slow.

Nevertheless, the Cu atoms dispersed at random in the crystal are in an unstable state, so a compromise occurs: the copper atoms gather together, but since they can only be displaced over small distances, the grains rich in copper are *very small* and *very numerous*. These are the GP zones, which in this case have dimensions of around 10 nm. Since these are regions in the alloy

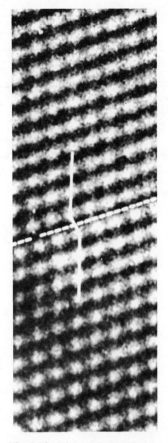

Fig. 5.16 High resolution electron micrograph of a stacking fault in a germanium crystal. (A. Bourret, Centre for Nuclear Studies, Grenoble.)

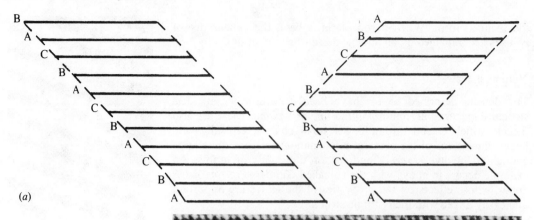

(a)

(b)

(c)

Fig. 5.17 Twinning in crystals. (a) Diagram showing the succession of layers in a crystal with perfect close-packing, and in a pair of twinned crystals. (b) High resolution electron micrograph in which individual atoms can be seen in a twinned arrangement similar to the second diagram of (a). The apparent reinforcement of the atoms in the twin plane is an interference effect. (A. Bourret, Centre for Nuclear Studies, Grenoble.) (c) Twinned natural crystals of chrysoberyl. (N. Bariand, Mineralogical Laboratory, University of P. and M. Curie, Paris.)

where the Cu concentration is abnormally high, the crystal lattice suffers great distortion (Fig. 5.18). So the zones are quite large defects interrupting the regularity of the crystal. Now the ductility of a metal, as we have seen, depends on the movement of

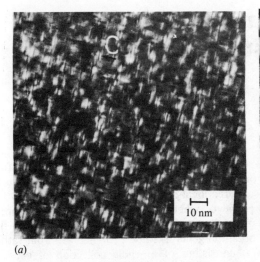

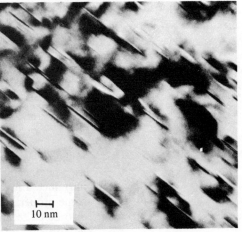

(a)   (b)

dislocations which allows easy slip between atomic planes. While a dislocation is certainly very mobile in a perfect crystal, its movements are blocked by zones of irregular structure, and GP zones, by their size and number, are particularly effective obstacles. The zones form spontaneously in the alloy when it is quenched to ordinary temperatures and this explains why the Al–Cu alloy hardens on ageing.

This example shows particularly well that too crude an approximation to the atomic structure of a solid is not sufficient to account for some of its interesting properties. In other words, it is not enough to know only the average or ideal structure of a crystal, but the defects must be known as well. Since some of these are quite subtle, it is clear that more refined methods of observation are needed and we then move to a more advanced state of the knowledge of matter. We give one more example.

A light alloy, as we have just seen, is soft after quenching but hardens after ageing at room temperature (and also by annealing at a moderate temperature). It is also well known that steel is hardened by quenching but that it becomes soft again after annealing or tempering. Presented in that way, the two sets of observations seem to be contradictory and it seems, too, that they just have to be accepted as peculiar experimental facts. However, if we start from the idea that a metal is hardened because of defects in the structure, the link between the two sets of facts is apparent: in both, *it is the existence of defects that strengthens the metal.* The difference arises from the particular way in which the defects are created in the two metals. In the light alloy, it is the 'zones' that are formed by diffusion through the solid during the ageing period. In the steel, a part of the metal changes its crystal structure during the quenching itself: the new phase, *martensite*, appears as needles with a lot of interstitial carbon: the metal around the martensite is thus highly distorted and these defects harden the metal. Tempering of the steel allows the structure to reorganize and thus to reduce the importance of defects, with a consequent increase in ductility.

Fig. 5.18 A crystal of Al–4% Cu alloy after quenching and ageing: (a) after 30 hours at 100 °C and (b) after 10 days at 130 °C. The electron micrographs show the zones where the agglomerations of copper atoms produce great distortions in the crystal lattice. (Yoshida, Cockayne and Whelan, Oxford.)

6 From crystals to crystalline solids

Hitherto, we have taken the crystal with its regular atomic structure as typifying the ordered state of matter, although we did have to introduce a few scattered local defects to account satisfactorily for some properties. However, when a pure substance changes from a disordered to an ordered state, for example by solidification of the liquid[1], what we obtain in general is not a *single crystal* but rather a *crystalline solid*, consisting of a whole collection of crystals (sometimes called *crystallites*). If there are many crystallites, we have a *polycrystalline solid*.

A model of a single crystal gives the positions of its atoms with respect to one of them taken as origin. When this origin is fixed at a given point, the *orientation* of the model then has to be specified—that is, the directions in space of certain characteristic rows of atoms called *axes*. If the model is turned about the atom at the origin, the directions of the axes are changed but clearly the *relative* positions of the atoms are unaltered.

When a pure liquid solidifies, the small crystals that grow in it will all have the same structure, but there is no reason why they should all have the same orientation. Therefore, in general, when the whole mass has solidified, the result is a *juxtaposition of crystallites of the same structure but with different orientations*. Such a 'polycrystal' is typical of the great majority of *crystalline solids* (i.e. those which are not glassy or amorphous).

The single crystal

In spite of what we have just said, however, it is sometimes possible for a homogeneous solid body to be composed of a *single crystal*, with dimensions ranging between a fraction of a millimetre and several centimetres or even decimetres. This would mean that the crystalline network had not only the same structure at every point but the same orientation as well. Such single crystals are encountered among natural minerals, such as the precious stones found in jewellery—diamond, sapphire or topaz, for example. Nowadays there are also many types of synthetic single crystals manufactured for industrial purposes—rubies, large quartz crystals, silicon, transparent crystals for optical uses, and so on (Fig. 6.1).

These single crystals, particularly those of the natural minerals, often have a distinctive external shape: they are bounded by plane

---

[1] The change could equally well be by condensation from the vapour or by crystallization from a solution.

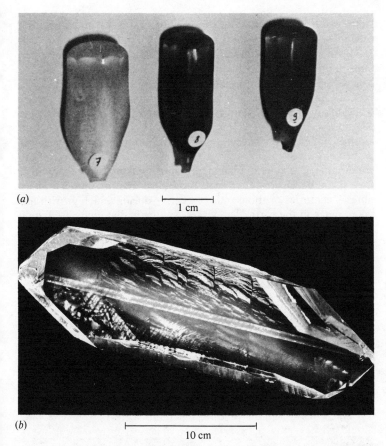

(a)

1 cm

(b)

10 cm

Fig. 6.1 (a) Synthetic crystals of corundum (Al_2O_3) produced by Verneuil's method (controlled cooling from the melt). The single crystals are produced without external faces. (Photographic library of the Palais de la Découverte.) (b) Synthetic quartz produced by growth from a small quartz needle, the seed immersed (controlled in a solution of silica at 500 °C and at a pressure of 1000 bars. (Industrial Nuclear Fuel Company, Annecy.)

faces making angles with each other whose values are very precisely characteristic of the given type. These are what we call **crystals** in the everyday use of the word. They were the subject of some of the earliest mineralogical studies which gave us a method of classifying crystals through observations of their shape. Today, we know that it is the structure on the atomic scale which characterizes the crystal and that the external shape is only a secondary property that may not even be very well defined. In fact, the external morphology has a direct relationship with the atomic structure (Chapter 4): the outside faces are always parallel to net planes of the lattice and so the angles between these faces are dependent on the geometry of the lattice. The natural faces of a crystal differ in their development because of anisotropy, which causes the speed of growth to vary according to direction.

Under what conditions can a single crystal begin to grow? Suppose, as an example, that it originates from a melted phase. Then there must be, right from the beginning of solidification, *only one* small crystal or *seed* in the liquid. This will then progressively grow, as atoms of the liquid come and attach themselves to the surface in a regular arrangement strictly in accordance with the atomic structure of the seed. The same structure must continue from layer to layer without any alteration and, in addition, no other seed must be allowed to start growing in the liquid.

Those are difficult conditions to bring about, so that it is easy to understand why the synthesis of large single crystals is a very delicate operation and one that is still far from being mastered. We do not know exactly why it is that some crystals grow quite easily while many other substances never yield single crystals above a certain microscopic size. Not only that, but some crystals found in nature cannot be produced at all in the laboratory. Sometimes this is because we cannot, of course, always achieve the proper conditions of temperature and pressure, but more than anything else it is because we cannot reproduce the enormously long times—centuries or thousands of years—during which growth has taken place in the earth. The largest synthetic diamonds are barely a millimetre across, whereas the largest known natural diamond, the 'Cullinan' from the Transvaal, measured 10 cm by 6 cm in its original state before cutting.

Crystal clusters

It is often found that large natural or synthetic crystals with quite well-defined external shapes occur not singly but in clusters, grouped together more or less irregularly. For instance, quartz prisms of various shapes and sizes can sometimes be seen jumbled up on the walls of underground caves known as geodes. Or, again, small calcite plates can be found grouped together in what is called a 'desert rose' because of its appearance (see Fig. 6.2).

Fig. 6.2 A crystal cluster of calcite ($CaCO_3$) forming a 'desert rose'. (Photographic library of the Palais de la Découverte.)

2 cm

The scale of such clusters varies enormously. Some are easily seen with the naked eye, while others can only be identified by mineralogists from their external form with the help of a microscope. Nowadays, the scanning electron microscope allows us to extend this method of identification down to very small crystals indeed. This instrument has a resolving power of the order of 0.1 μm and it provides pictures of surfaces of specimens even when they are quite uneven. Figure 6.3 is an example showing the quite striking relief that can be obtained in such pictures.

Some mineral crystals are frequently found in pairs, joined together along a certain plane and having fixed relative orientations. These are the *twinned crystals* we have already discussed in the last chapter: an example was given on p. 120.

Polycrystals or polycrystalline materials

Polycrystals form by far the majority of crystalline solids, including all the pieces of metal we normally use, all the rocks found in the earth, and so on. They all consist of a collection of small single crystal grains, or crystallites, jammed one against the other *without any gaps between them*. There are two differences between this class of material and the one described in the last section. The first difference is qualitative: in polycrystals, the elementary grains are generally quite small, often microscopic or even sometimes not visible under the optical microscope. The second difference is more

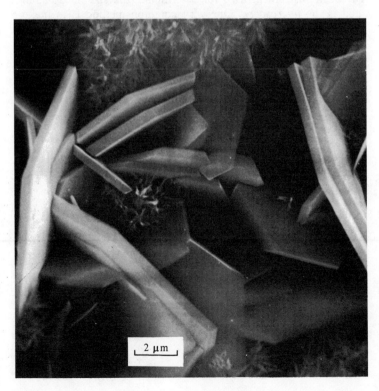

Fig. 6.3 Microcrystals under a scanning electron microscope identifiable by their shape: hexagonal platelets of calcium aluminosulphate and fine needles of hydrated calcium silicate. (M. Regourd, Centre d'Études et de Recherches sur les Liants Hydrauliques, Paris.)

2 μm

fundamental: the grains in polycrystals are not bounded by faces having any relation to the atomic structure, as Fig. 6.4 shows.

In many rocks and metallic alloys, the crystal grains are of several different substances (Fig. 6.5). Firstly, however, we consider only those cases in which all crystallites have the same chemical composition and the same atomic structure, as in a pure substance. A given sample is then characterized by the size and shape of the elementary crystals or grains, and by the orientation of their axes relative to the solid as a whole or relative to each other. All these factors taken together define what is called the **texture** of the solid. (Take care to distinguish between the *texture* of a solid, where the element is the small crystal grain, and the *structure* of a crystal, where the element is the atom, ion or molecule.)

The properties of a given solid are determined by both texture and structure: in addition, we saw in Chapter 5 that the ideal crystal structure must be supplemented by a description of the defects that occur in a real material. This last point is the most troublesome one of all because the experimental data are rarely precise or complete.

The texture of a solid has a considerable effect on its properties, and we shall encounter a striking example of this later on p. 134. It is clearly very important to be able to determine this texture and we now discuss how this is done.

The experimental determination of the texture of solids

The basic instrument used for determining the **size** of grains in a solid is the *microscope*. One very common method is to make a section of the solid and polish it to produce a perfectly flat surface.

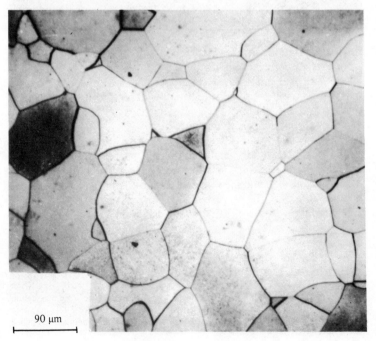

Fig. 6.4 The polished surface of a polycrystalline metal, aluminium, under an optical microscope. The grains of various shapes have dimensions of the order of 0.1 mm. (P. Lacombe, Metallurgical Laboratory, University of Paris-Sud.)

90 μm

felspar

mica

quartz

This is then treated with chemical agents which attack it: the choice of agent depends on the nature of the solid and on what one wants to see. Some chemicals, when applied to metals, etch the boundary lines between crystallites, while others etch the surfaces to produce variations of light and shade from one grain to another according to their orientation. Another method, suitable when dealing with rocks, is to prepare thin sections or platelets through which light can pass. The grains in such a specimen act on *polarized* light passing through them, but the action is different for different thicknesses and orientations. The instrument used for this technique is thus called a *polarizing microscope.*

Electron microscopes[1] extend the range of observations beyond those possible with optical instruments, largely because of their much greater resolving power. They enable grains much smaller than 1 μm to be examined, and in addition can detect dislocations in the crystals. They provide the best available method for evaluating the degree of perfection of the grains.

To obtain the greatest amount of information about the **orientation** of the crystal grains, *X-ray diffraction patterns* are used. However, this method by no means gives the orientation of the axes of each individual grain (with grains of about 10 μm in size

Fig. 6.5 A thin section of granite, 1/50 mm thick, under a polarizing microscope. The rock is formed from crystals of quartz, mica and felspar, distinguished from each other by their optical properties. (Photographic library of the Palais de la Découverte.)

[1] The electron microscope referred to here is of a different type from the scanning electron microscope mentioned previously for examining surfaces. The instrument here is one that passes electrons *through* the thin sections and is thus often called a transmission electron microscope.

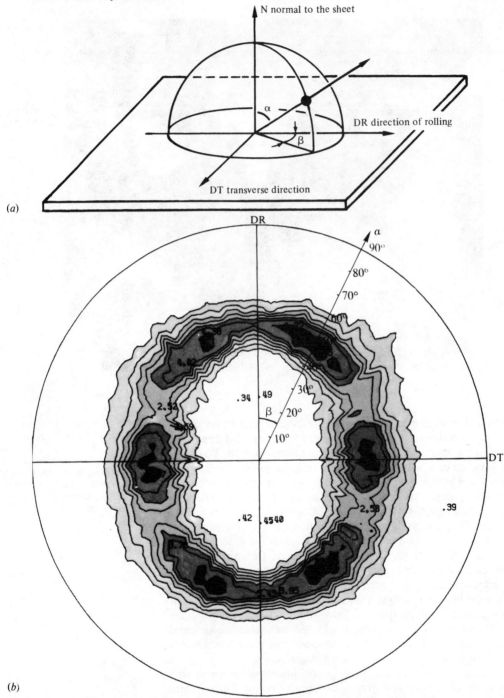

(a)

(b)

Fig. 6.6   Determination of the texture of a sheet of rolled steel. (a) The direction of a microcrystal axis is defined by its 'colatitude' α and its 'longitude' β with respect to the direction of rolling. (b) A pole-figure obtained from X-ray diffraction. It is a map giving the distribution of the axial directions of a family of microcrystals as a function of α and β. The family shown here have axes parallel to the sides of the cubic unit cell of the crystals. If the sheet were isotropic (i.e. with the same properties in every direction) the distribution would be uniform because the axes would be oriented at random. The pole-figure thus indicates *preferred orientation*. (P. Parnière, Iron and Steel Industry Research Institute, St.-Germain-en-Laye.)

there would be one million of them per cubic millimetre!). The significant information revealed by the X-ray patterns is the *statistical distribution* of the directions of the axes for the whole collection of crystallites.

In the simplest cases, the directions of the axes are distributed quite randomly: *the complete solid is then isotropic even though each crystal grain is anisotropic.* The anisotropic properties simply cancel out on average. However, in many solids there are some quite well-defined directions of the axes which are more favoured than others. The whole material is then no longer isotropic and the grains are said to exhibit *preferred orientation*. Figure 6.6 shows the results of determining the anisotropic texture of a rolled steel sheet by this method. Using the normal to the sheet as a reference, it reveals how the directions of the cubic unit cell axes are distributed. Instead of being uniformly spread over the diagram, they are clearly concentrated in certain particular directions.

Grain boundaries and their distinctive role

The assumption that the properties of a polycrystalline solid are an average of those of the individual crystallites is not generally a very good approximation to the truth. This is because each grain interacts with its neighbours and because the boundaries themselves have their own properties. As the two-dimensional diagram of Fig. 6.7 shows, the regions between two crystals with different orientations consist of layers of atoms in irregular positions because their surroundings are not uniform as they would be in the interior of a grain. We now know that the irregularities are confined to a thin layer of only a few atomic diameters across. The number of atoms involved is thus extremely small but, as we have already seen in the case of crystal defects, that does not by itself prevent some properties from being greatly affected. As an example, *a metal with small grains has a greater mechanical strength than the same metal with large grains.* The deformation of a metal under the action of an external force is due to the movement of dislocations, and such movement is impeded by the disordered layers at the grain boundaries. The smaller the grains, the larger is the total area of the boundaries and the greater is this impedance. Thus, for a given force, the deformation is reduced and the metal is tougher.

In some alloys a particular kind of deformation occurs in which the change of shape before fracture under quite a weak force is much greater than normal. This occurs at higher temperatures and is sometimes known as *viscous flow* even though the material is solid. It is also known as *superplasticity*. The crystal grains slide over each other as if they were hard particles whose surfaces were coated with a viscous layer. The boundaries between the particles contain those atoms which, as Fig. 6.7 shows, are slightly disordered and less uniformly packed than in the inside of a normal crystal. The sliding occurs most readily at temperatures near the melting point and the deforming of a metal using this process is thus known as *hot-working*[1] (Fig. 6.8).

[1] The softness of lead and tin is due to the same effect. Their low melting points mean that 'hot-working' can occur at room temperatures.

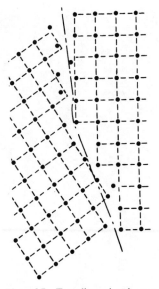

Fig. 6.7 Two-dimensional representation of the grain boundary between two crystals with different orientations. The disordered atoms are confined to a thin layer whose thickness is of the order of one or two atomic diameters.

Fig. 6.8 A 60 cm diameter hemi-sphere of an alloy of titanium, aluminium (6%) and vanadium (4%) produced by hot-working at 920 °C. The operation lasts two hours and is only half as costly as the usual technique. (B. Baudelet, Faculty of Science, Metz.)

60 cm

Another consequence of the atomic disorder at the interface between adjoining grains is that the diffusion of foreign atoms is made easier: the boundaries become *preferred sites* for the collection of impurities. If these impurities should reduce the cohesion between grains, the metal can become very fragile. Some chemical agents also attack the boundaries preferentially because the atoms there are less tightly bound than those in the interior of a crystal. Grain boundaries can then be the starting points for bad corrosion (Fig. 6.9).

The effect of anisotropy in the texture

The importance of the anisotropy of metallic textures in various industrial manufacturing processes can be illustrated by a few examples.

Consider the thin sheet of steel whose anisotropic texture was demonstrated in Fig. 6.6. It is found that the mechanical strength is not the same for stresses applied lengthways (i.e. parallel to the direction of rolling) as for those applied in the transverse direction. Not only that, but the plastic deformation of the sheet is up to three times greater in its own plane than in the direction of its normal.

Again, consider a thin circular sheet transformed into a cup by pressing (also called deep-drawing)—that is, by exerting a very strong pressure on the sheet between a punch and a die. If the texture of the sheet is anisotropic, then the object produced may have an irregular shape (see the formation of 'ears' in Fig. 6.10). In the same way, the body of a car, which is also produced by pressing, will only have a satisfactory appearance if the texture is up to the right standard.

In other cases, a very anisotropic texture is actually preferred. Sheets of silicon–iron alloy used in transformers have better

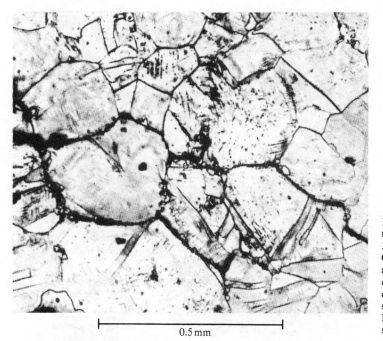

0.5 mm

Fig. 6.9 Grains in a polycrystalline material (brass: 70% copper, 30% zinc) beginning to fall apart. Corrosion along the grain boundaries, showing up as black lines, considerably reduces the cohesion of the grains and with it the strength of the metal. (P. Lacombe, Metallurgical Laboratory, University of Paris-Sud.)

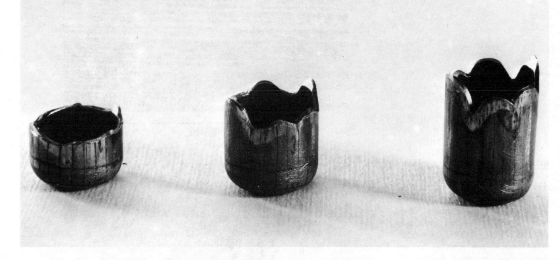

magnetic properties when the crystallites have an orientation which is fixed in relation to the magnetic field, so it is obviously desirable to make the proportion of well-oriented grains as large as possible.

The mechanical and thermal treatment given to a metal of fixed composition determines its texture, and thus provides metallurgists with various methods for improving it. These can be the bases for worthwhile reductions in costs. However, the treatments must be carried out with precise control of the conditions and with strict monitoring of the texture using microscopes and X-ray diffraction.

Fig. 6.10 A circular disc of sheet-steel pressed into a cup by deep-drawing between a punch and die. The sheet has a very pronounced anisotropic texture which produces irregular deformations around the upper edge called 'earing'. (P. Parnière, Iron and Steel Industry Research Institute, St.-Germain-en-Laye.)

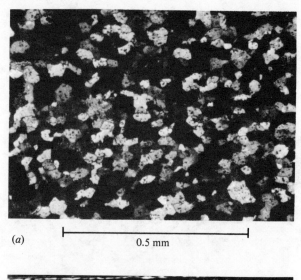

(a)   0.5 mm

(b)   0.5 mm

Fig. 6.11 (a) Micrograph of a metal, aluminium, after annealing showing well-formed crystal grains. The different shades are due to variations in orientation. (b) Micrograph of the same metal after work-hardening. The individual crystals are too small to be visible. The streaks are parallel to the direction of stretching during rolling and show the preferred orientation of the crystals. (c) The metal in (b) under the much greater magnification of the electron microscope. Areas of about 0.5 to 1 μm across appear to be separated by regions with a high concentration of defects which play the part of wide grain boundaries. (P. Lacombe, Metallurgical Laboratory, University of Paris-Sud.)

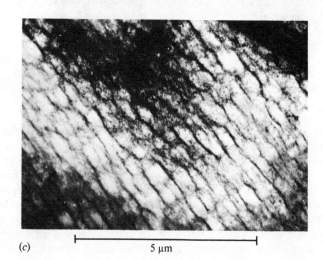

(c)   5 μm

Work-hardening and annealing (or recrystallization) in metals

In the industrial production of a metal, its first solid form is the ingot, obtained by pouring the melt into a mould and allowing it to cool. The ingots are subsequently formed into successively thinner sheets, or into bars, by rolling. Finally, the bars can be reduced in cross-section even further to make rods or even wires of less than one millimetre diameter by drawing them through dies.

In the initial ingot, the metal is polycrystalline and the grains are often quite large—sometimes of the order of one millimetre or more. As the metal is rolled into sheets or bars, the large crystals break up into smaller ones which remain stuck together in a coherent mass. Ultimately, however, in the rods and thin wires, the crystals become so small that they are no longer visible under the optical microscope (Fig. 6.11(b)), although X-ray diffraction patterns show that they still exist and possess the normal atomic structure of the metal. When viewed under the electron microscope (Fig. 6.11(c)), the grain boundaries cannot be clearly picked out: uniform areas *less than 1 μm across* are separated by highly imperfect regions which are much wider than the boundaries between the well-formed grains we have met previously.

In addition, X-ray diffraction shows that the small crystals formed from the original ones no longer have the same orientation as they had at the start, but now have directions depending on the treatment received. For instance, in the case of a drawn copper wire, the diagonals of the cubic unit cells are all directed along the axis of the wire, or at any rate no more than a few degrees from it (Fig. 6.12). Thus, the only fixed direction is that of the unit-cell diagonal, so that the crystals take up any orientation as long as it merely involves a rotation of the unit cell about that direction. This is called *fibrous texture*.

Finally, the cold-working of the metal by rolling or drawing creates a *large number of imperfections* in the interior of the crystals, most of them being *dislocations*.

After all this treatment, the metal is said to be **work-hardened**, a state which has the following characteristics: very small crystallites (only a few tens of nanometres across) with an extremely anisotropic orientation (fibrous texture being only one example) containing an enormous number of defects, mainly dislocations.

The effect of work-hardening is to give the metal *great mechanical strength*. Take, for instance, a work-hardened copper rod, 5 mm in diameter and 50 cm long, and support an 800 g weight at its end as shown in Fig. 6.13(a). It merely bends a little and the bending is elastic (this means that the rod would straighten out again if the weight were removed, and that giving the end a small displacement and releasing it would cause it to vibrate).

The work-hardened condition is not stable, however. Keeping the rod at 500 °C for a few minutes is sufficient to modify its state considerably. If the same weight is then attached to the end of the 'reheated' rod, it bends permanently and to a much greater extent than before (Fig. 6.13(b)). From being tough and strong the copper has become quite soft. A metal in this latter state is said to be **annealed** or recrystallized.

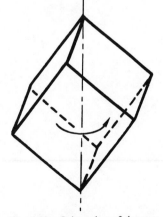

Fig. 6.12 Orientation of the crystalline unit cell in a metal after drawing. The diagonal of the cube is directed along the direction of drawing, which will be the axis of the rod or wire. All orientations about that direction are possible.

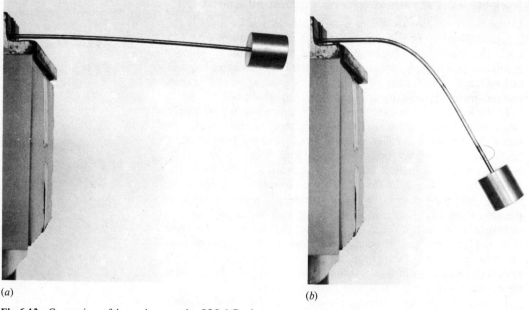

(a)   *(b)*

Fig. 6.13 Comparison of the mechanical properties of a copper rod which has been: (a) work-hardened and (b) annealed. In (a) the copper is strong and elastic; in (b) it is easily deformed.

At 500 °C, the atoms must possess quite a high mobility for the new crystals to be formed from the small imperfect ones of the work-hardened stage. Annealing changes the texture of a metal but not its atomic structure (the copper rod of Fig. 6.13 has, in both states, crystals with a face-centred cubic unit cell of side 0.363 nm). The recrystallization produces well-formed grains easily visible under the optical microscope (Fig. 6.11(a)) at least 0.1 mm across, and the fibrous texture of the work-hardened state has disappeared. Moreover, the new crystals are not so imperfect, the number of dislocations having been considerably reduced (by a factor of about 10 000).

The simple experiment of Fig. 6.13 shows very clearly the great influence of the texture of a polycrystalline solid on its macroscopic properties.

Solids crystallizing in more than one phase

The solids considered so far have consisted of crystals which all had the same structure and differed only in their sizes, shapes and orientations. However, that is clearly a special case and, more generally, a crystalline solid will consist of several 'phases', each a collection of crystallites with their own structure. The grains of a given phase will still, as before, differ from each other in their orientation.

A complete description of a solid thus involves the crystal structures of each phase as well as their textures. The latter can be extremely complex, for we must specify not only the shapes and orientations *within* each phase but the relationships between the grains of different phases as well, such as their relative positions and orientations. All of these will play a role in determining the properties of the solid.

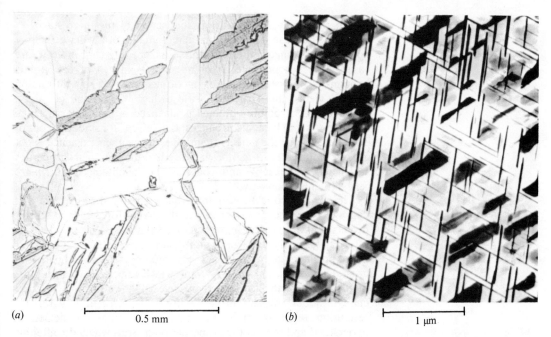

(a) |———————————————|
             0.5 mm

(b) |———————————————|
              1 μm

Figure 6.14 gives two typical examples of the textures of metallic alloys containing two phases, A and B, say. In the first, the crystallites of phase B (α-iron), which are quite small, are mixed with the larger grains of phase A (γ-iron). They have quite irregular shapes and orientations. In the second example, the crystallites of phase B are thin platelets inside a large grain of phase A. The platelets are parallel to one of three perpendicular planes which are the faces of the unit cell of A.

The physical properties of solids with several phases are not an average of those of the separate phases, but are dependent on the particular texture, just as they are in the case of a single phase. It is the interaction between adjacent grains that is important and it is found that *the properties of a composite material may be better than those of any of its constituents*. We shall be returning to this point in Chapter 10 when we deal with composite materials in general. Multiphase crystalline solids are only a particular case, and amorphous solids can be involved as well.

Fig. 6.14 Examples of alloys with several solid phases. (a) Steel with two phases (α-iron and γ-iron) having crystals of comparable size. This is an optical micrograph: the two phases can be distinguished because of their different appearance after chemical etching of the polished surface. (b) Electron micrograph of an alloy with composition 90% aluminium and 10% silver. The white background is a crystal of aluminium, while the dark lines and patches are grains of a phase Al_2Ag in the form of very thin platelets (10 nm thick, 0.5 μm across). The platelets are parallel to the three faces of the cubic unit cell of aluminium. Two of the systems are seen in section and the third is in the plane of the photograph. (P. Lacombe, Metallurgical Laboratory, University of Paris-Sud.)

Summary

We end this chapter by summarizing the stages in the construction of a model of a crystalline solid. In Chapter 4 we described the structure of a geometrically perfect crystal. Then, in the next chapter, we found more precisely that all real 'good' crystals departed a little from perfection[1] because of scattered defects or local imperfections. Finally, in this chapter, we have described the texture of solids, that is, how they are composed of collections of small crystallites.

[1] We shall see in Chapter 10 that there are also very 'poor' crystals in which defects are very numerous.

Experimentally, the texture is the easiest of the three stages to determine because we can obtain direct images of it with the help of optical, and sometimes electron, microscopes. Except in very rare cases, it is not possible to obtain direct images of the atomic structure of crystals, but for 'good' crystals X-ray diffraction always yields the structure with a high degree of certainty. However, the structure obtained is idealized in that it doesn't take account of small local disturbances: the defects are smoothed out of existence. As for the defects themselves, while their existence is not in any doubt, there is no way of determining their properties completely and exactly, and what data can be obtained on them are very difficult to come by.

The general conclusion is thus that we know the structure of a completely crystalline solid very well, except for details about the defects that are present in every crystal but which are too dilute to affect appreciably the overall atomic arrangement. This 'exception' may appear trivial but is in fact very important. Many properties of crystalline solids, particularly some which affect our everyday use of them, are controlled by irregularities that we do not yet know enough about. Much current research is directed towards the resolution of these difficulties so that we may be helped to understand and predict the properties of crystals which do, after all, constitute a large part of the solids all around us.

7 Pure liquids, liquid mixtures and solutions

After our long exploration of the crystalline state, we now return to the disordered state of matter and, in particular, to liquids. We start by recapitulating some of their characteristics which were established in Chapter 2.

Since it is possible to pass continuously from gas to liquid, there can be *no sharp discontinuity in structure between the two states*, where by 'structure' we mean the arrangement of molecules. Just as a gas has a disordered structure, so has a liquid. The essential difference between gases and liquids is that the density of the latter group is so much greater and is close to that of solids. The number of molecules per unit volume does not differ by more than about 10% between the two condensed states. It follows that, if neighbouring molecules in *crystals* are *in contact*, then the situation must be much the same in *liquids*.

Pure liquids

Vaporization and boiling

When we say that we observe a certain material to be a liquid, we are in fact observing one of the constituents of a *two*-phase system in equilibrium: one is the liquid and the other its vapour. According to the equilibrium diagram of Fig. 2.1, the pressure of the vapour, p, is fixed at each temperature below the critical point. The simplest case is that of a pure substance contained in a vessel whose volume lies between two values: that of the material completely in the liquid state and that when it is all saturated vapour at the temperature of the experiment. The proportions of liquid and vapour adjust themselves by vaporization or condensation until the vapour pressure equals p. If the vessel also contains air, the equilibrium conditions of the liquid and its vapour remain unchanged: they are independent of the total pressure P, whatever it may be. The partial pressure of the vapour is still p, and hence that of the air is $P - p$.

If now the liquid is open to the atmosphere, the conditions in the immediate neighbourhood of the surface of the liquid are still the same (as we saw on p. 27): the total pressure P is one atmosphere (1.013 × 10⁵ Pa), and the partial pressure of the air at the surface is (one atmosphere $- p$), where p is the saturated vapour pressure. It follows that the system can only be in equilibrium if p is less than one atmosphere. Take water as an example: when open to the air and heated, its temperature rises until it reaches 100 °C where *p* attains a value of one atmosphere. The water then *boils*, its temperature remaining fixed at 100 °C while it changes violently into vapour.

Liquid nitrogen can only exist in equilibrium with its vapour alone below −147 °C, which is its critical temperature (p. 24). However, when open to the atmosphere, it can only be in equilibrium below −196 °C, the temperature at which its saturated vapour pressure is one atmosphere. These temperatures are far below room temperatures, so that the vessel containing such a liquid must be thermally insulated. However, no matter how good the insulation, if the outside of the vessel is at room temperature, the nitrogen will eventually rise to −196 °C, its boiling point under atmospheric pressure. At this point, the heat received from the outside through the vessel cannot produce a rise in the temperature of the liquid but is used instead to change it into a vapour. The loss of liquid in this way can be greatly reduced by improving the thermal insulation of the container, as in a dewar flask (Fig. 7.1). In practice, the loss of liquid nitrogen from the largest storage tanks is calculated to be less than 0.5% per day, while the losses from large tankers used for transporting liquid methane at its boiling point of −161 °C are about 0.1% per day.

Limits to our knowledge of the structure of a disordered substance

A liquid, being both *dense* and *disordered*, seems to exhibit two contradictory features. The molecules cannot be situated just anywhere, because they are quite closely packed and cannot penetrate each other. It is therefore quite certain that the disorder in liquids is not complete, as it is in gases at low pressure, and therein lies one of the essential problems of their structure. This problem has been approached from two directions: firstly from experimental data and secondly by constructing theoretical models that are continually adjusted to bring them into agreement with the facts of observation.

Only X-ray or neutron diffraction can provide direct information about the structure of liquids just as they do for crystals, except that the results obtained with disordered arrangements yield much less information than those with strictly periodic structures.

For the moment, we shall consider the simplest case of pure liquids with monatomic molecules, such as the inert gases or metals. In order to describe the structure of such a liquid exactly, the positions of the atomic centres would need to be determined with an uncertainty that is small compared with the diameters. Neither X-rays nor neutrons can yield a true image of the structure, not only because the movement of the atoms is too rapid in relation to the duration of the diffraction experiments, but also because such an outcome is unfortunately completely outside the theoretical capabilities of the method.

In general, what can be obtained *at best* is *a statistical view of the surroundings of an atom*. In a simple crystal (say with one atom per unit cell), each atom has the same surroundings in accordance with the periodicity of the crystal. The 'average structure' that is

a   evacuation outlet
b   evacuated space
c   liquid nitrogen
d   adsorbent to improve
     the vacuum

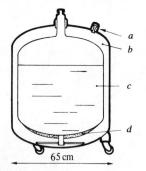

65 cm

Fig. 7.1   Cross-section of a 100 litre dewar flask containing liquid nitrogen. The two walls of the vessel are of metal with the space between them evacuated to a pressure of about 1 Pa.

obtained is then none other than the structure itself. In disordered matter, however, many details are lost in the 'average structure' which is all that experiment will give us.

Let us imagine that a minute creature, gifted with such sharp sight that it can distinguish a separation of 0.01 nm, is situated on an atom of the liquid. It looks around and notes the directions and distances of all the neighbouring atoms at a given instant of time. Our 'demon' then carries out the same observations sitting on a large number of different atoms. Finally, it takes the average of all its observations and hands them to us. Since a liquid is statistically isotropic, *the average picture of the structure will have spherical symmetry*. This means that the only significant parameter we need to consider is *the distance from the atom at the origin of our observations, r*.

The first thing we know is that, since the atoms are impenetrable, there can be no centres of other atoms at a distance away that is less than one atomic diameter. Furthermore, at large distances (say at more than five or ten diameters), the influence of the atom at the origin will be felt less and less, so that the average distribution of atoms seen from the origin tends to become practically uniform. More exactly, if there are on average n atoms per unit volume in the whole collection, there will be $n \, dv$ atoms in a small volume dv provided that its distance from the origin is greater than ten diameters. As we get nearer to the origin, the packing of the atoms as if they were hard spheres causes a variation in their density which can be written as a function of the distance r. The mean number of atoms in a small volume dv can then be written as $P(r) \, n \, dv$, where $P(r)$, known as the *radial distribution function*, is generally different from unity. *It is this distribution function and this alone which is determined by X-rays or neutrons*.

Short-range order in liquids

We have just argued, in effect, that $P(r)$ is zero for $r < a$, the atomic diameter, and tends to unity at large distances. In intermediate regions, $P(r)$ oscillates around unity: at some distances there are more atoms, at others less, than if the distribution were uniform. This is what is meant by *short-range order*, which is a characteristic feature of condensed disordered structures. However, at large distances, liquids are disordered (no long-range order): in other words, if the distance between two atoms exceeds about ten diameters, there is no longer any correlation between the positions of the pair. This is in complete contrast to the existence of order even at very large distances which is characteristic of a perfect crystal.

However, the nearer atoms do not have just *any* arrangement. One atom is surrounded by a layer of nearest neighbours following a certain pattern which is not strictly unvarying as in a crystal but is only roughly so, showing differences from one atom to another. Between the atom at the origin and the *second* layer of neighbours, the possible variations increase, and so on, in such a way that beyond about the fifth layer, disorder has become almost complete.

Figure 7.2 shows the distribution function $P(r)$ for liquid mercury as determined from X-ray diffraction. If we consider first a powder consisting of small mercury crystals whose orientations are absolutely random, the whole specimen is isotropic even though each small crystal is not. A distribution function $P(r)$ can be defined for such a powder just as for a liquid, and it would show a series of sharp peaks situated at the various interatomic distances in crystalline mercury. (In fact, these distances fluctuate slightly around their theoretical values because of thermal vibrations, thus broadening the peaks a little.) If these sharp peaks are now compared with the $P(r)$ curve for liquid mercury, it is striking to see that the first of them coincides almost exactly with the first maximum in the continuous curve for the liquid.

This result suggests that *the surroundings of an atom in the liquid strongly resemble those of an atom in the corresponding crystal.*

This idea provided the basis for the earliest models of liquid structure: a collection of crystals, originally assumed to be very small (about 1 to 2 nm) and later on to be capable of great distortion—a somewhat vague hypothesis which led to theories with wider scope. However, the model is now abandoned because it did not explain some of the more detailed experimental data yielded by X-rays. Moreover, the existence of distinct crystals, even if very imperfect, brought their boundaries into the picture. Such heterogeneous zones separating the more homogeneous regions in the crystals themselves do not seem to be consistent with many liquid properties.

The hard sphere model: irregular close-packed arrangements

At present, the most promising geometrical model is based on *irregular close-packing of spheres* (or, in a similar way to the h.c.p.

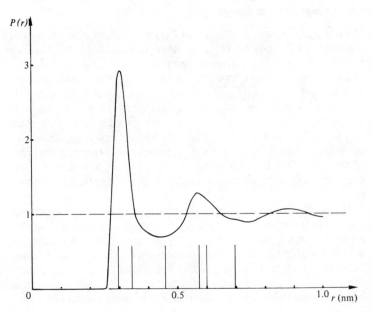

Fig. 7.2 Radial distribution function $P(r)$ for atoms in liquid mercury. The vertical lines show the distances of an atom from its neighbours in crystalline mercury. It is clear that the most probable interatomic distances in the liquid correspond to the interatomic distances in the solid.

and f.c.c. structures in Chapter 4, on an irregular array of tetrahedra). Atoms are represented by hard spheres as in crystalline structures and they are packed in such a way as to touch each other (Fig. 7.3). Around a single sphere it is possible to place as many as twelve others which all touch it, but there is no room for a thirteenth, even though the twelve do leave a little space between them so that they can be moved slightly while still maintaining contact with the central sphere (the experiment is easy to carry out with table-tennis balls). As a result of the spaces which allow the movement, a variety of arrangements is possible in which the whole thirteen spheres keep the same average density.

We have already met two of the possibilities in the h.c.p. and f.c.c. structures. As can be seen from Fig. 4.41, in these cases the twelve outside spheres form eight triangles and six squares, but while the triangles are as closely packed as they can be, there is a small space at the centre of each square: it is there that the spaces for movement are to be found among the twelve spheres. Another possibility is to distribute the extra spaces uniformly, when the spheres form only identical equilateral triangles but are not in contact—the distance between their centres is 1.01 diameters. The centres of the spheres are then at the twelve corners of a regular polyhedron with twenty faces called an *icosahedron*. The spheres can be grouped in opposing pairs, each pair defining one of six five-fold axes (that is, rotation about such an axis by multiples of $2\pi/5$ brings the polyhedron into a position identical with its starting position). The black patches on a football form an icosahedron.

However, a five-fold axis is incompatible with a regular crystalline array (p. 67) and it is impossible to make icosahedra grow by using one as a motif to fill space by periodic juxtapositions, as can be done with a cubic or hexagonal motif. If a new layer of spheres is added to the model, each sphere touching or nearly touching three of the previous layer to form tetrahedra, large holes necessarily occur but these are still not large enough to accommodate another sphere. The whole structure cannot there-fore be close-packed everywhere. Even if we try to make a heap of spheres as compact as possible, its structure cannot be regular unless it starts from a small initial h.c.p. or f.c.c. arrangement which can grow by following the chosen scheme.

The existence of disordered close-packed structures rests on the purely geometrical fact that a central sphere *cannot be completely covered* by twelve touching spheres of equal diameter. To convince ourselves of this, we need only consider the two-dimensional arrangement of spheres in a plane (Fig. 4.38). In this case, one sphere is *exactly* surrounded by six other touching spheres, yielding a pattern with six-fold symmetry. Such a motif can grow *regularly* and *indefinitely* to produce a hexagonal layer which is the arrangement automatically taken up by spherical balls that are assembled closely together in a plane (as in the Bragg bubble raft, Fig. 5.6): they can be said to 'crystallize'. So, in two dimensions, there is *no disordered* close-packed structure.

However, when spherical balls are packed into a bag, they always take up an *irregular* three-dimensional close-packed

Fig. 7.3 An irregular close-packed assembly of identical spheres. Numerous groups of four touching spheres form regular tetrahedra, but it is impossible to arrange them all in this way.

arrangement. A regular or 'crystalline' array cannot be produced, either by pressure or by shaking to give the balls a little mobility.

If the number of spheres packed irregularly like this in three dimensions is large enough, the average density is always less than that of the close-packed crystalline arrangement, which, as we saw on p. 93, was 0.74 (that is, the ratio of the volume of the spheres themselves to the total occupied volume—in fact, the value is $\pi/3\sqrt{2}$). It is a remarkable fact—and one that has only recently been theoretically explained—that when spheres are packed as tightly as possible in an *irregular* three-dimensional arrangement, there is found *in practice* to be a well-defined upper limit to the density. The ratio of the volume of the spheres to the total occupied volume is found to be 0.64, which is 14% less than the value of 0.74 for the crystalline case. This constant ratio provides justification for the measurement of quantities of grain by the volume occupied as well as by the mass, a practice that has been in use in the cereal trade from time immemorial. (The word 'bushel' has its origin in the name of the container used in measuring corn.)

Excellent confirmation of the above model is provided by the inert gases, whose density decreases by almost exactly 14% on melting. In metals, however, the decrease is much less at only about 5%. To predict such a figure using the same general model, it would be necessary to assume that a high proportion of the liquid has a quasi-crystalline structure, which does not agree with X-ray data. Our conclusion must therefore be that metallic atoms do not behave exactly like hard spheres: the ions in metals do not make contact through their complete outer electron shells but are maintained at their equilibrium distance by the free electron cloud. It is quite possible, therefore, that the distance between ions could decrease slightly when the arrangement changes from the crystalline form to the liquid state, so giving them a smaller effective diameter and offsetting the normal density change from 0.74 to 0.64.

Detailed analyses have been made of the contacts between spheres in real examples of irregular close-packing (Fig. 7.4). Numerous groups with pentagonal symmetry are observed as well as many regular tetrahedra of four spheres in contact. There are also, as could be predicted, many local irregularities with quite varied forms.

A much more precise way of proceeding is to 'construct' a heap of 1000 or more spherical balls by computer and to calculate the distribution function for pairs for any given arrangement. With sufficiently large numbers, it is found that the function coming out of the calculation is very similar to those given by X-rays for liquids. The model is thus in agreement with experiment. However, it would be going too far to say that its validity is thereby established, because some quite different models also yield distribution functions that are close to the experimental ones.

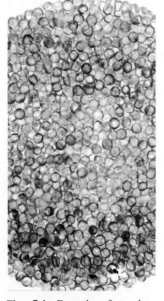

Fig. 7.4 Example of a close-packed assembly of hard spheres: small glass balls in a crystallizing dish.

The structure of molecular liquids

Liquids in which the 'particle' is a single atom are exceptional. True, there are also fused salts such as NaCl, where the particles are Na^+ and Cl^- ions, but in the great majority of liquids the particle is a fairly complex molecule, in general without spherical symmetry. It is much more difficult to establish a model in these cases, because a pair of molecules is not then defined only by the distance between their centres but also by their relative orientations. These two quantities cannot be obtained separately if the only available experimental data are those yielded by X-ray diffraction, so that all construction of new models relies on a certain number of unverifiable hypotheses.

When two molecules with elongated ellipsoidal shapes come into contact, their minimum distance apart varies according to the angle between their major axes: we sense that they would have a certain tendency to align themselves parallel to each other but we are left with a very qualitative description. Grains of rice packed in a bag (Fig. 7.5) provide a rough picture of the structure of a liquid with elongated molecules.

There is also another complication to be considered. Models involving close-packed arrangements rely solely on the idea that neighbouring molecules attract each other but with completely *undirected* bonds. That, however, is only a particular case because, just as in crystals, *non-close-packed structures with directed bonds* can exist in liquids. As an example, liquid silicon contains atoms that are surrounded by fewer neighbours than those in a liquid metal. As another example, but one that is most important, we shall look in detail at the structure of water.

The structure of water
The structure of an isolated water molecule is well known (Fig. 7.6). The oxygen atom is linked to two hydrogen atoms (more exactly to two protons) at a distance of 0.097 nm along directions making an angle of 104.5° with each other. The bonds are made by contributions from the valence electrons of oxygen and the electrons of the two hydrogens.

The structure of ice is also known exactly (Fig. 7.7). A given molecule is surrounded by four others whose oxygen atoms are arranged as are silicon atoms, each at the vertex of a regular tetrahedron at a distance of 0.276 nm from the central one. Each oxygen is also surrounded by four hydrogens on the O—O lines, but two of them are nearer the central O than are the other two, forming a H_2O group which is only slightly different from the isolated molecule (the H╲O╱H angle is 110° instead of 104.5°). The other two hydrogens (protons) are linked to the oxygen by hydrogen bonds (see Fig. 4.27(b)). Ice has a very open structure: the volume of the isolated water molecule shown in Fig. 7.6 is 0.0168 nm^3 and so the fraction of the volume occupied by molecules in an ice crystal is only 0.55, compared with the 0.74 for regular close-packing of spherical atoms.

When ice melts, some ordered molecular arrangements similar to those in the crystalline form persist in the liquid state, but the molecules are no longer *all* linked to neighbours as if by fixed 'struts'. Some of them in fact draw closer together in an irregular way, making a more compact configuration: *that is why the density*

Fig. 7.5 A layer of rice grains lying in a close-packed formation on a flat plate. Notice how numerous small groups of parallel grains are formed spontaneously.

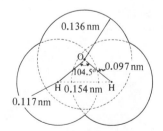

Fig. 7.6 The structure of the water molecule. The spheres traced around the O and H nuclei have the Van der Waals radii of the atoms (p. 14).

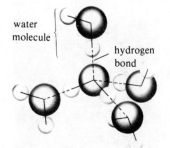

water molecule

hydrogen bond

Fig. 7.7   The basic motif of an ice crystal is formed by a water molecule surrounded by four others at the corners of a regular tetrahedron. Two neighbouring oxygen atoms are linked by hydrogen bonds. (L. Salem, *Molécule la Merveilleuse*, Intereditions, 1979.)

of water is greater than that of ice, against the general rule for the melting of a crystal with a close-packed structure. The anomalous behaviour of water is thus the result of the open structure of ice. When the temperature rises from 0 °C, the structure of water becomes more and more unlike that of ice and so its compactness increases. Just above 0 °C, this tendency outweighs the normal thermal expansion that occurs in all liquids and it explains the maximum in the density of water at 4 °C. The fact that the volume of a given mass of water increases on freezing is extremely important in determining the conditions of life on earth, principally because ice floats on the oceans. There are also other anomalies in the physical properties of water which stem from its molecular structure.

Thermal motion

As we have previously seen in Chapter 2 (p. 28), the atoms of a liquid, as of all forms of matter, are undergoing perpetual movement which becomes faster as the temperature rises. In gases, molecules move rapidly along linear paths broken only by their collisions with other molecules. The 'mean free path' is of the order of several tens of nanometres, so that a molecule diffuses over macroscopic distances in quite short times (about 6 mm per second).

In crystalline solids, however, atoms remain near points that are fixed by the regular structure of the crystal. Their movements are restricted to vibrations around these fixed points because of the restoring forces provided by the action of numerous near neighbours. In practice, no atoms could change position unless there were a few missing ones (vacancies, p. 107) to allow a certain amount of mobility. Diffusion is always very slow and only occurs appreciably at high temperatures (about 1 μm per hour at 800 °C).

Liquids are an intermediate case: a given molecule is caged in by its neighbours and oscillates in this more or less well-defined and restricted space but, because of irregularities in the structure, openings occur quite frequently in the 'cage' and so, in the course of normal movement, the central molecule may escape into a neighbouring space where the process starts again. Diffusion of the molecule will occur at a speed that is *slow in relation to a gas but fast in relation to a solid*. It is in this way that the static model of liquid structure must be completed by the dynamic aspect. Atoms are packed closely together but are in perpetual motion; the surroundings of a given atom are constantly being modified and some atoms are changing their positions, yet overall the same average density is always retained.

Liquid cohesion and surface tension

A quantity of liquid has *no shape of its own*: we say that a portion of matter is a liquid when its external surface can be very easily deformed by the action of small impulses. We are going to try to understand this behaviour using the model of its atomic structure.

When a quantity of liquid is deformed, it remains all in one piece and its volume stays constant. The cohesive forces that exist between the molecules keep them in contact and prevent them from becoming dispersed. These forces are of the same type as those we have examined in explaining the cohesion of crystals: in liquids consisting of molecules or of inert gas atoms, they are *Van der Waals forces* (p. 75). Sometimes, in addition, the *hydrogen bond* is involved (p. 77), particularly with organic molecules. There are also *electrostatic interactions* in fused ionic salts such as NaCl because the 'particles' in this case are Na^+ and Cl^- ions. Finally, in molten metals, just as in the crystalline form, there are *metallic bonds* formed by the free valence electrons, so that a molten metal is also a good electrical conductor.

The thermal motion of atoms in the liquid state is too great for the binding forces to maintain the molecules in a permanent and well-ordered structure as they do below the melting point. A close-packed coherent assemblage is formed but the molecules are not prevented from sliding over each other. To return to our billiard-ball model, it is as if they were magnetized, so that a cluster of molecules can change shape while remaining stuck together in a compact group.

The equilibrium shape of a quantity of liquid is determined by two factors: the existence of *surface tension* and the action of *gravity*, only the first being an intrinsic property of the liquid.

Every molecule at the surface of a liquid is subject to the attention of its neighbours but, unlike one in the body of the liquid, all the forces are directed towards one side only. There is thus a resultant directed towards the interior (Fig. 7.8) and this gives rise to a pressure round the exterior as if the liquid were surrounded by a *stretched elastic membrane* which naturally tends to reduce its area. The effect of this surface tension is that a liquid tends to take that shape which, for a given volume, has the minimum surface area: the sphere. Table 7.1 gives several examples of solids having the same volume but different shapes and shows that all have surface areas larger than that of a sphere.

When a liquid touches a solid wall, the molecules at the surface are subjected to the action of the liquid on one side and of the solid on the other. Depending on the nature of the particular substances, the liquid may be *attracted* by the solid (*wetting* it) or *repelled* by it. Thus, a drop of water spreads out on a very clean glass plate but retains its drop-like shape if the plate is greasy. These contact forces are the cause of **capillarity**.

Table 7.1 Relative surface areas of bodies with the same volume but different shapes.

Nature of the solid	Volume	Surface area	Surface area per unit volume
Cube of side a	a^3	$6a^2$	6
Regular tetrahedron of side b	$(\sqrt{2}/12)b^3$	$\sqrt{3}b^2$	7.20
Cylinder of diameter and height c	$(\pi/4)c^3$	$(3/2)\pi c^2$	5.54
Sphere of diameter d	$(\pi/6)d^3$	πd^2	4.84

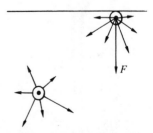

Fig. 7.8 A molecule is bound to its neighbours by Van der Waals forces, say. When the molecule is near the surface, these forces have a resultant F directed normal to the surface and towards the interior.

Liquids are *weighty* because we normally handle them in the earth's gravitational field. Just as a movable solid is in equilibrium when its centre of mass is as low as possible, so the equilibrium shape of a quantity of liquid is that which has as low a centre of mass as possible, taking account of any constraints on it. For that reason, a liquid occupies the bottom of the vessel that contains it and is limited by an *upper horizontal free surface*. Although this behaviour is considered to be characteristic of a liquid, it is, in reality, due to the combination of two factors: one arises from the structure of the liquid and is its deformability; the other arises from external circumstances and is the existence of the gravitational field.

In practice, the shape of the liquid is determined by the combined effect of the two forces, weight and surface tension, although in most cases it would be better to say that it was determined by the weight and modified slightly by feeble surface effects, because of the relative magnitudes. Thus the free surface of a liquid is not strictly horizontal but is bounded at its edges by a *meniscus* which may be either convex or concave. In the same way, when a very fine-bore capillary tube is dipped into water, the liquid rises in the tube to a level above the outside one, as in Fig. 7.9(a). The formation of a drop at the end of a tube as in Fig. 7.9(b) is the result of the opposing actions of weight and surface tension.

If the weight is absent, as in the interior of a satellite, only the surface forces remain. The phenomena which are then observed differ greatly from our everyday experience. For example, a quantity of liquid will float in the air and take on a spherical shape, while in a vessel the shape depends on any walls which it wets (Fig. 7.10). Under weightless conditions, the classical experiment on the condensation of a liquid at its saturated vapour pressure does not show the liquid forming at the bottom of the vessel but rather in a cloud of droplets which continually grow by combining with each other.

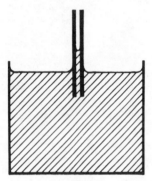

(a)

(b)

Fig. 7.9 Examples of capillarity. (a) Rise of water in a fine-bore tube (3 cm for water in a tube of 1 mm diameter). (b) Formation of a drop at the end of a dropping-tube.

The viscosity of liquids

Having studied the equilibrium of a liquid, we now look at its non-equilibrium behaviour or its *flow*, a process of continual deformation while remaining in one piece. The molecules are therefore still in contact and slide over each other in a movement which is accompanied by friction. The greater the friction, the more the liquid is said to be *viscous* (or the less it is fluid).

We take as an example the motion of a liquid through a tube of circular section. A characteristic feature of this is that the speed varies over the cross-section of the tube from the edge to the centre. Those molecules in contact with the walls adhere to the solid surface and remain stationary. Away from the walls, the speed increases with distance and attains a maximum value along the tube axis. Coaxial cylindrical layers of the liquid thus have different speeds and this gives rise to the friction associated with viscosity. An analogous analysis was made in the case of gases on p. 49.

Experimentally, the magnitude of the viscosity is defined by a coefficient which can be obtained by measuring the outflow from a straight tube of known length and diameter in which the liquid is made to move by a known pressure difference. It is intuitively clear that the greater the viscosity of the liquid, the less is the outflow from the tube. Another method is to measure the speed of fall of a ball in the liquid. The fall is rapid when the viscosity is low, as in ether for example, but is much slower in a thick oil.

Can the coefficient of viscosity be *calculated* from a knowledge of the structure of the liquid alone? Although this has been done for a perfect gas, *it is impossible for a liquid* and it is interesting to analyse why this should be so. In a gas, the 'friction' between layers with different speeds was due to a transfer of momentum from one layer to the other. The calculation needed only a single molecular parameter, its effective cross-section, because there was no inter-action between molecules except by collisions whose consequences could be easily evaluated. By contrast, the molecules of a liquid are in contact, so that their friction depends on their external structure and on short-range interactions that we do not know how to calculate. It is made all the more difficult by the disorder which makes contacts very irregular.

In spite of that, several correlations observed in liquid viscosities (see Table 2.2) can be simply explained. The least viscous liquids are those having small molecules with disordered close-packed arrangements. This is because there is no way in which one molecule can 'hook' on to another: even in water, where there are hydrogen bonds, the viscosity remains of the same order since such bonds break very easily and re-form with another molecule so that, overall, there is very little resistance to movement.

However, very viscous liquids such as oils and glycerine have enormous molecules with complex shapes and, most important of all, quite a few points with intermolecular hydrogen bonds. In another example, liquid silica is very viscous because the great majority of oxygen atoms are linked by very strong covalent bonds to two silicons, even though the atoms do not occupy regular positions as in a crystal.

In a general way, liquid viscosity falls when the temperature rises: in glycerine, for example, it decreases 2.5 times between 20 °C and 30 °C. The reason is that the vibrations and rotations of the molecules increase with temperature and lower the chances of 'hooking'. Although the analogy is not very good, the similar case of a powder of small grains comes to mind: when agitated, it can be made to flow quite easily, a method used for the transport of solids.

(a)

(b)

Fig. 7.10 Experiments on liquid equilibrium when enclosed in a spherical vessel: (a) under normal conditions, the liquid partially fills the vessel with a free horizontal surface while (b) under weightless conditions, the liquid (which wets the walls) adheres to the vessel and leaves an empty space at the centre. (J.-L. Boulay, Office National d'Études et Recherches Aéro-spatiales, Chatillon-sous-Bagneux, France.)

Liquid mixtures

From the moment they are placed in contact, any two gases that do not react chemically will mix together to form a single *homogeneous gas* whatever the relative proportions. The properties of the mixture can easily be predicted because each gas behaves as if it were alone, strictly so if they are perfect gases, and approximately so if they are not too imperfect.

At the other extreme, two crystalline solids which do not react have structures that are so well determined and so rigid that it is normally impossible for them to fuse together: they can only be placed in close contact without mutual penetration. There is only the exceptional case of certain metals when they have very flat and *very clean* surfaces which do fuse when they are pressed strongly together at a high temperature. The atoms of one diffuse into the other to form a layer with a thickness of several hundredths of a millimetre which consists of a solid solution (p. 95) or of an intermediate compound.

Liquids, in this area as in others, have a behaviour that lies between these two quite distinct cases (no mixing in solids, complete mixing in gases), and one that is, as usual, much more complicated.

The experimental facts

Keep in mind that we are considering pairs of liquids that do not react chemically.

There are numerous liquids which *mix together in all proportions*. If they are undisturbed, the mixing occurs quite slowly yet spontaneously, but agitation speeds up the process considerably. The result is a perfectly homogeneous liquid: a well-known example is that of alcohol and water.

However, there are pairs of liquids which are *completely immiscible*. For instance, when oil is poured on water the lighter oil floats on the surface and forms a layer containing no water, while underneath is a layer of water containing no oil. Even after violent shaking, these two layers re-form. If two such liquids have the same density, for example silicone oil and salt water, the oil floats in the water as if it were weightless because the Archimedean upthrust exactly balances the weight. The shape taken up by the oil then depends only on the surface tension and on the form of any walls which it wets (Fig. 7.11).

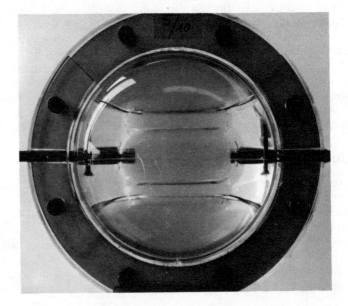

Fig. 7.11 Behaviour of two immiscible liquids of the same density (silicone oil and coloured salt water). The equilibrium configuration depends on the surface tensions and the interfacial tensions with the small metal plates placed in the inside of the vessel. (Experiment carried out by the European Propulsion Society to simulate the behaviour of the fuel in the tank of a satellite under weightless conditions.)

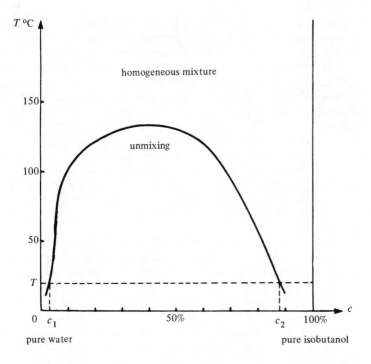

Fig. 7.12 Equilibrium diagram of a water–isobutanol mixture. Any state defined by temperature T and concentration of the alcohol c which lies completely above the curve represents a homogeneous solution. Below the curve, there is unmixing: that is, coexistence of two phases with concentrations c_1 and c_2, the limiting homogeneous solutions at the temperature T.

Finally, there can be *partial unmixing*. This is the case with a mixture of water and isobutyl alcohol (isobutanol) (Fig. 7.12). If the alcohol is added to the water at a temperature of 20 °C, the mixture is homogeneous up to a concentration of 4% by weight. When water is added to the alcohol, the limit of homogeneous mixing is 11% water, 89% alcohol. For intermediate concentrations, the liquid forms two separate layers, corresponding respectively to the two limiting solutions with 4% and 89% alcohol. When the temperature rises, the limiting concentrations approach each other until above 130 °C there is complete miscibility in all proportions.

Interpretation of the behaviour of liquid mixtures

Can the behaviour of liquid mixtures be predicted, or even just explained, in terms of the molecular structure of the constituents? In this case, as in so many others, a complete and quantitative answer is entirely outside our present capabilities, simply because it is liquids that are involved. We must therefore be content with qualitative ideas but these, limited as they are, do provide ways of understanding some of the observations we can make in everyday life.

The basic idea is that it is possible to replace many molecules in a liquid by other different ones without destroying the structure since it is a disordered one. (This is unlike a crystalline solid with its rigid structure, where such substitution is usually only very limited, although even here there are solid solutions (p. 95).) Substitution in a liquid is tolerated even more when the two molecules have very similar shapes and sizes and, above all, when they have similar attractive forces between them (that is, of the same nature and of comparable strengths). Under these conditions it is easy to see how a homogeneous mixture of the two types of molecule could occur.

If each molecule has a much greater affinity for its own kind than for the other, it can only tolerate a low proportion of different molecules, or even none, among its nearest neighbours. There is then either complete immiscibility or at least a limit to concentration of possible homogeneous mixtures. These limiting mixtures coexist without disturbing each other and produce *partial unmixing*. When the temperature rises, disorder in the structure increases because of the thermal motion of the molecules and, in general, the limits to the solubility relax since a molecule can accept a greater number of different ones in its immediate environment. In some cases, there is even complete miscibility above a certain temperature.

The key to the problem of predicting the degree of affinity between molecules with known structures would be a precise quantitative knowledge of their interactions. Unfortunately, such knowledge is still at a quite rudimentary stage, although a few conclusions, some of them almost commonsense, can be drawn. For example, it is hardly surprising that mercury atoms, with their metallic bonds produced by free electrons, and water molecules, with their hydrogen bonds, form liquids that are completely immiscible.

However, water does have an affinity for *polar* liquids like itself, that is, with molecules containing electric dipoles having equal and opposite +ve and −ve charges, such as the O−H or the C=O radicals. Such dipoles play an important role in the interaction between polar molecules through their electrostatic fields and give them a common character. That is why water mixes in all proportions with ethyl alcohol (ethanol, C_2H_5OH). Pentanol ($C_4H_9CH_2OH$) by contrast is almost insoluble in water in spite of its polar character, a difference produced entirely by the very great disparity in size between the small molecule of water and the much larger one of pentanol.

The importance of the polar nature of the molecules is further demonstrated by the complete miscibility between two *non-polar* liquids such as benzene (C_6H_6) and hexane (C_6H_{14}), whereas there is no miscibility between water, which is polar, and benzene, which is non-polar, or between water and hydrocarbons such as petrol (gasoline) for example.

Solution of a gas in a liquid

There is clearly little similarity between the structures of two such substances that are to be brought together. The gas is mixed with the vapour of the liquid, whose partial pressure is fixed and is dependent only on the temperature of the system. In addition, the gas can dissolve in the liquid up to a limited concentration which, for a given gas–liquid pair, is proportional to the pressure of the gas. Those are the experimentally observed facts. The limiting solubility of the gas varies greatly from one system to another: for example, under normal conditions of temperature and pressure, it is practically zero for argon in mercury, very low for oxygen in water (5 mg/100 g), greater for CO_2 in water (0.2 g/100 g) and very large for NH_3 in water (60 g/100 g).

It is easy to conjecture what will be the structure of a gas dissolved in a liquid. All the molecules are in a disordered state, so that those of the gas must be interspersed among those of the liquid in a completely irregular way. Their concentration is likely to be all the greater when there is greater affinity between the two types of molecule: thus an inert gas atom has no special bond with water molecules since only weak Van der Waals forces (p. 75) are involved and the low solubility is therefore entirely predictable. With CO_2 or NH_3, the attraction between the molecule of the gas and that of water is the result of what is virtually a chemical reaction; in effect carbonate or ammonium ions are formed and the solubility is very high.

We now take a look at the equilibrium in a given gas–liquid system as illustrated in Fig. 7.13. The existence of equilibrium does not mean that no interchange of molecules takes place between the gas and liquid, but merely that there is exact compensation between the numbers escaping from the liquid and those returning to it. The rate at which gas molecules *enter* the liquid is proportional to the numbers hitting the surface and hence to the density or pressure P of the gas. The rate at which gas molecules *escape* from the liquid increases when there are more of them dissolved per unit volume and is thus proportional to the concentration of dissolved gas, c. The equality of the two rates of movement in opposite directions means that *the solubility of the gas in the liquid is proportional to the gas pressure, P.* This is, in fact, the experimental law, expressed as $c = kP$.

The gas above the wine in a corked bottle of champagne at 10 °C contains CO_2 at a pressure of about three times atmospheric, while the wine contains 5 g of dissolved CO_2 per litre. When the bottle is uncorked, the CO_2 pressure falls, so that the dissolved gas thus has an excessive concentration and must escape: this accounts for the small bubbles that give champagne its sparkle. Equilibrium is established quite slowly because the CO_2 is, as we mentioned above, combined with the water and also because CO_2 is denser than air and remains on the surface to form a layer in which the partial pressure of CO_2 is high. Only after shaking the wine and after the CO_2 has diffused into the atmosphere does the wine become 'flat'.

Very often, it is only possible to understand certain observations when it is realized that *equilibrium is not being achieved.* Thus, in a large mass of calm water, the oxygen from the air effectively saturates the surface layers, but the deeper ones have little or no dissolved oxygen. Homogeneity can only be established by diffusion of oxygen through the liquid and we have seen that, although this is possible, it is very slow, particularly if the water is very deep. However, if the water is stirred up, the amount of oxygen dissolved throughout the whole mass is much greater. That

gas G (pressure P) + vapour L (pressure p)

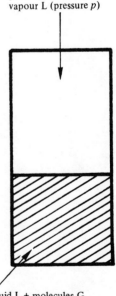

liquid L + molecules G at a concentration $c = kP$

Fig. 7.13 Solution of a gas in a liquid.

is why the conditions of life for a fish, needing oxygen to survive, are so different in a calm pond and in a turbulent stream. It is also why the water in an aquarium has to be aerated by bubbling air through the tank: it would be insufficient to rely on the oxygen from the air in contact with the free surface.

Solution of a solid in a liquid

Experience tells us that when any solid is immersed in a liquid, it dissolves until the concentration of the solution reaches a 'limit of solubility', a quantity that varies greatly from one solid–liquid pair to another. Table 7.2 gives values for the limiting solubility of several common solids in water.

When we consider the part played by water over the whole globe—through rain, rivers and oceans—it is clear that the degree of solubility in water is a vital property of all materials. Natural minerals on the earth's surface have not disappeared, for instance, because they are not very soluble, yet natural waters are never pure: they are more or less dilute solutions of the substances with which they have been in contact. *Evian water* contains only 0.3 g of dissolved salts per litre, but some mineral waters are more potent (5 to 8 g per litre) while sea water contains mainly NaCl at a concentration of about 3.5 g per litre.

Aqueous solutions play other important roles: they are indispensable elements in the physiology of all beings, animal or vegetable, and they are involved in the majority of the processes in the chemical industry and in many chemical techniques.

Judging by these various examples, matter in the form of solutions is an essential part of our life and activities. A qualitative model of their behaviour can be given without difficulty but, as is usual with condensed disordered matter, details of the structure are too complex to yield a precise quantitative theory or to enable predictions to be made of all their physical properties.

Table 7.2   Solubility limits of several solids in water at ordinary temperatures (values given in g dissolved by 100 g of pure water).

Level of solubility	Example	Solubility limit
Very soluble	Caesium acetate	945
	Calcium iodide	750
	Antimony chloride	600
	Sugar	204
	Silver nitrate	120
Soluble	Sodium chloride	35
	Mercuric chloride	7
	Gypsum	0.24
Almost insoluble	Calcium carbonate	1.5×10^{-3}
	Mercurous chloride	0.2×10^{-3}
	Barium sulphate	0.2×10^{-3}
	Lead carbonate	0.1×10^{-3}

The dissolving of solids and the structure of solutions

Let us consider a system consisting of a solid immersed in a liquid, water for example. As we have so often remarked, the molecules in both phases (or, more generally in the solid, the constituent 'particles') are always in motion. Some will leave the solid surface and float off into the liquid while others at the same time will leave the liquid to be deposited on the solid (note that we are talking in both cases about the particles of the solid). When these two processes compensate each other exactly, there will be equilibrium. Now the number of molecules captured by the solid will be proportional to the number of collisions they make on the surface, so it is natural to assume that the flux from liquid to solid would be proportional to the concentration of dissolved molecules. Against that, the flux from solid to liquid has a fixed value because the solid structure is unchanging and because the concentration of the solution is not in general high enough to affect the situation. As the solid dissolves, its concentration in the liquid increases, so increasing the liquid-to-solid flux: until, that is, the two fluxes are equal when equilibrium will be established for a certain concentration of the solution, for the given substances, and at a given temperature. This limiting concentration is the solubility limit of Table 7.2.

To try to account for the variations between the solubilities of different pairs of substances in contact, we must first describe the structure of a solution, limiting ourselves to the most important case of aqueous solutions.

The solubility of molecular crystals

The first case we examine is that of solids consisting of molecular crystals. The 'particle' of such a crystal is a molecule which passes into the liquid unchanged. Thus, when sugar dissolves in water, the sugar molecules are dispersed, each being surrounded by water molecules. This can only occur if there are bonds between the solvent and the dissolved molecules with a strength at least comparable with that between the molecules of the solid. The transfer of a molecule from the solid to the liquid is then not opposed and dissolving can take place. If, however, the attraction between the solvent and the molecules of the solid is much weaker, any molecule which by chance entered the liquid would rapidly return to the solid through the much stronger attraction. The solubility limit would then be very low or the solid practically insoluble.

In the table of solubilities in water (Table 7.2), there occurs once again the distinction we found for liquid mixtures between *polar* and *non-polar* molecules (see the definition on p. 77). Water is strongly polar because of its O—H groups and so it readily forms bonds with other polar molecules. Crystals whose molecules include O—H, C=O radicals, etc., are the most soluble, while those which have only CH_2 and CH_3 radicals, such as the paraffins, have no affinity for water molecules and are not soluble.

Even where there are bonds to nearby water molecules, a solute molecule with a shape and size very different from those of water would necessarily produce distortion in the arrangement of layers immediately surrounding it in the liquid. It is natural to suppose that the weaker are these distortions, the greater would be the solubility, and it is in fact true that, in a series of homologous molecules, solids with smaller molecular weights are more soluble.

The solubility of ionic crystals

Apart from metals, the other class of solids is that in which there are no molecules in the crystal but in which the atoms are linked by covalent or ionic bonds (Chapter 4). *Purely covalent crystals are insoluble*, since to detach an atom from the crystal would need the severance of a very strong bond with the expenditure of a lot of energy which could not be recovered through linking the atom to a water molecule.

In complete contrast, *ionic crystals* can be soluble or even very soluble. The ions with opposite charges are bound by electrostatic forces in the solid, but they can be dispersed in water on being brought into contact with it: what is called *electrolytic dissociation* is in fact just such a dispersion. For example, the NaCl molecule does not exist since the crystal is only a regular and close-packed assembly of alternating Na^+ and Cl^- ions. In water, the attractive force between the charges of opposite sign is greatly reduced in comparison with the force between the same charges at the same distance in a vacuum. (The Coulomb force between two charges Q and Q' at distance r apart is $F = QQ'/4\pi\varepsilon_r\varepsilon_0 r^2$, where ε_r is the relative permittivity of the medium. Its value for water is about 80 which implies a corresponding decrease in the value of the force between the ions in water.) The ions thus have much less tendency to reunite.

There is, however, another important feature here. When in water, ions do not remain 'naked'. The electric field that their charge produces around them attracts the electric dipoles of the water molecules, which then form layers surrounding the ions. The effect is particularly marked with small ions which produce a stronger field with the same charge. So the ions, being protected by their layers of water molecules, cannot get near enough to each other for the attractive force to become very strong and they remain dispersed in solution.

The hydration of ions is very convincingly demonstrated by the following facts. The energy needed to detach single pairs of K^+ and Cl^- ions from a KCl crystal is 685 kJ mole^{-1} (see p. 80), whereas to dissolve a KCl crystal in water needs only 12 kJ mole^{-1} in the form of heat. This means that nearly all the energy needed to disperse the ions is recovered by the formation of hydrated ions in the solution. They are therefore very stable because their binding energy (that is, of the ions plus their entourages of water molecules) is of the same order of magnitude as that for an ionic crystal. The structure of hydrated ions is not in general very well known except in a few simple cases. We do know, for instance, that the Cr^{3+} ion is surrounded by six water molecules at the corners of a regular octahedron (Fig. 7.14).

An ionic solution is not, therefore, simply water containing ions: the structure of pure water is quite seriously perturbed by both the electric and geometrical effects of the ions, particularly when they are significantly larger than the water molecules. The first layer of water molecules round an ion is strongly bound by electrostatic forces but, outside that, between the hydrated ion and normal water, there is a region in which the molecules are less ordered because of the different constraints at the inner and outer boundaries of the region.

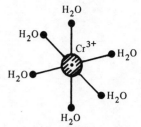

Fig. 7.14 Diagram of the Cr^{3+} ion hydrated with six water molecules.

Physical properties of solutions

The structural differences between pure water and solutions, whether ionic or molecular, make us realize that the latter, even if dilute, could possess some properties that deviate from those of water itself.

For example, the *volume of the solution* is not in general merely equal to the sum of the volumes of the water and the dissolved solid. This is both because the volume occupied by the dissolved molecules or ions is not equal to their volume in the original solid and because the water does not have its normal density owing to the disturbances in its structure near the dispersed particles. An exceptional case of this, and a very curious one, can be quoted: when 8 g of magnesium sulphate, having a volume of 3 cm³ in the dry state, is added to 1000 g of water, with a volume of 1001.35 cm³ at 18 °C, the salt dissolves and produces a solution whose total volume is less than that of the water alone (Fig. 7.15).

The *viscosity* also changes because the friction between molecules is not the same for pure water as for solutions, which contain regions where the structure is perturbed by dissolved molecules or ions.

Ionic solutions, or *electrolytes*, are *conductors of electricity*. The electric charge is carried by the ions of opposite signs which move under the action of an electric field created when a potential difference is established between two electrodes. The conductivity of the electrolyte depends on the resistance to their motion encountered by the ions. However, given the complex forms of the hydrated ions, the perturbations they produce in the structure of water, and the electrostatic interactions, it is easy to appreciate that a rigorous theory would be difficult to formulate. In fact, although much research work has been devoted to electrolytes since the introduction of the idea of an ion at the end of the last century, we still cannot account exactly for the conductivities of various electrolytes and their variations with temperature, concentration, and so on.

Crystallization

Suppose we have a saturated solution in equilibrium and open to the atmosphere so that the solvent is left to evaporate. As time

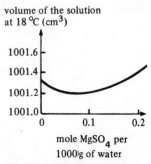

Fig. 7.15 The volume of a solution of $MgSO_4$ in 1000 g of water and its variation with the amount of salt dissolved. (Lewis and Randall, *Thermodynamics*, McGraw-Hill.)

goes on, the concentration of the dissolved substance increases and eventually exceeds the solubility limit. To re-establish equilibrium, a certain quantity of the solid has to separate out and is *deposited*. This is the way in which sea salt is produced by evaporation from salt marshes for example. The process is called *crystallization* and is the inverse of dissolution.

Crystallization can be achieved by another method: in general, solubility decreases when the temperature falls, so that, if the temperature of a saturated solution is lowered, the solid is deposited until the concentration of the solution returns to its limiting value at the new temperature.

By one or other of these processes, a situation is created in which dissolved molecules have a greater concentration than the liquid can tolerate: as soon as there is a chance encounter between two of the excess molecules as they move around there is a tendency for them to stick together and to remain so, in order to reduce their interactions with the solvent molecules. Other dissolved molecules then join the first two and so this small collection continues to grow, always by haphazard encounters. Eventually it could take on the form of a small crystallite of the dissolved substance: a sort of embryonic crystal. We postulate such effects by 'chance', but they are not really so improbable when we remember that the configuration of a collection of molecules changes at the rate of 10^{12} times a second. Thus, even in a time that is short on our time scale, phenomena can appear whose probability is very low.

If the small embryonic crystal *is* formed, it has a good chance of redissolving, but it can also grow by the arrival of new molecules taking their place in the regular structure of the crystal. If it is given the chance to reach a certain 'critical size', which can be theoretically evaluated, the embryo becomes a stable **seed**: not only can it no longer disappear, but it tends to grow more and more. In this way, a whole collection of small crystals is formed which constitutes the new-born solid phase. Its mass increases until the concentration of the solution is reduced to the solubility limit.

Certain points in a solution are more likely to be sources of seeds than others: irregularities in the walls of the container, dust particles immersed in the liquid, and so on. If the number of seeds which form simultaneously is very large, the total volume of deposited solid will be divided between great numbers of very small crystals. In general, these will be separated from each other so that a fine powder is obtained, but they can also stick to each other, giving a polycrystalline mass. If, with the same quantity of deposited solid, the seeds are less numerous, the grains will be larger and facets of individual crystals will then be seen with the naked eye. Finally, if we can succeed in having only one seed, a single crystal is obtained which grows with plane faces because of the anisotropy of the speeds of development.

This last method is one of those used in manufacturing synthetic crystals. A saturated solution is maintained strictly at a constant temperature to within 0.01 °C and the formation of seeds is avoided by taking great care to purify the solution and work with very clean apparatus. A small crystal of the dissolved substance,

Fig. 7.16 Growth of a crystal of ADP (ammonium dihydrogen phosphate) by very slow crystallization from a solution. The rod at the centre carries the initial seed and rotates in the liquid which is maintained at a temperature constant to within about 0.01 °C. The crystal grows with plane faces and reaches a diameter of 10 cm and a mass of 2 kg in approximately two months. (P. Chapelle, University of Paris-Sud.)

serving as the single seed, is suspended in the body of the solution which is gently stirred to keep the concentration uniform. The seed is rotated slowly and continuously at a few revolutions per minute (Fig. 7.16). Evaporation is controlled and is very slow, and in a few weeks crystals of several centimetres across are obtained without defects. This is the method used to prepare large crystals of ammonium dihydrogen phosphate (ADP) from which piezo-electric devices are sometimes made. It is also used to prepare quartz crystals from solutions of silica in an aqueous solution of caustic soda at a temperature of 500 °C and a pressure of 1000 bar. The crystals produced (Fig. 6.1(b)) are purer and more perfect than most natural quartz crystals and are used in the electronic and watch industries.

8 Non-crystalline solids: the amorphous or glassy state

This chapter deals with solids that belong to the disordered state of matter. In particular, we shall be concerned with their structure and with properties that stem directly from it. Such substances have two main characteristics. The first is that they are undoubtedly solids to those who see them, handle them and make use of them. The second is a negative one: it is that X-ray diffraction, being capable of revealing atomic structure, indicates that there are no crystals present.

The most logical name for these materials would thus seem to be *non-crystalline solids*, although, in fact, the terms **amorphous substances** or **glasses** are often used, and there is no generally accepted convention that distinguishes one of these names from the other. The word 'amorphous' (derived from the Greek 'without shape') has sometimes been used in a different sense: a rock consisting of microcrystals of calcite ($CaCO_3$) has been called 'amorphous limestone' by some geologists because of its indefinite external shape in contrast to a rock in which the individual calcite crystals are large enough for plane faces to be visible. To the physicist, both are crystalline solids: they have the same crystal *structure* (that of calcite) and are only distinguishable by their crystalline *texture* (Chapter 6).

Macroscopic characteristics of an amorphous solid

A substance normally exists in one of three states, crystalline, liquid or gaseous, depending on the temperature and pressure. It does not usually pass through an amorphous solid phase at all, so that in this sense it is an anomaly. Yet there are numerous examples to be seen around us, not even counting materials that have structures *intermediate* between crystalline and amorphous (the subject of Chapter 9).

The example we meet most frequently is what we commonly call *glass*. Hard and difficult to deform, even with quite strong forces, its shape is not easily changed without fracturing, yet it is quite easy to break. Such a substance is said to be brittle. If we examine it with the naked eye, or even with a powerful microscope, it appears to be a *continuous* homogeneous material. This characteristic is important because, as we shall see, it is the macroscopic equivalent of its atomic structure. Glass is transparent, for instance, because there is no irregularity to interrupt or deviate any rays of light that pass through it. When it breaks, the new surface that is created has a complicated shape but is always smooth (it is

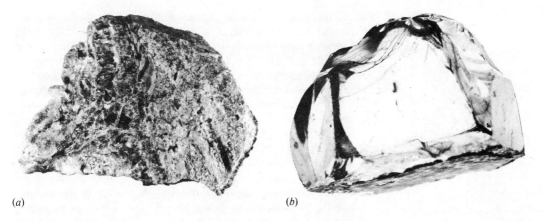

(a)   (b)

called a conchoidal fracture), whereas the fracture of a piece of metal or rock creates an irregular surface showing crystalline grains all stuck together (Fig. 8.1). Lastly, glass is *isotropic*, which means that no direction in its structure is different from any other, and this is also characteristic of all amorphous solids.

Fig. 8.1 (a) Fracture of a poly-crystalline solid: the surface revealed by the break is very irregular and contains a multitude of small facets having varied orientations. (b) Conchoidal fracture of an amorphous substance (glass): the new surface is broad and *smooth* with a slowly-varying curvature.

The transformation from liquid to amorphous solid

To examine how a substance changes from a liquid to an amorphous solid, we shall first analyse in greater detail than hitherto what happens in the normal case of a liquid transforming to a crystalline solid as the temperature falls through the melting point T_f. In such a case, as we have already argued, the transformation cannot be continuous because there is a fundamental discontinuity between disorder and order that has to be crossed.

Just as in the case of crystallization from a solution (Chapter 7), a *seed* having the structure of the crystal must first be formed in the cooling liquid. In the solution, the dissolved molecules are quite dispersed and need a large displacement to collect around such a seed; in the melt, however, the molecules are effectively as closely packed as in a crystal and so the formation of a seed only needs small displacements of the molecules: all that happens is a change in their local configuration.

During the never-ending thermal motion of the molecules, a few of them (several dozen, perhaps) may take up, quite by chance, the configuration of a crystal, forming an 'embryo' in a certain minute region. As long as the temperature is above the melting point T_f, the embryo would have only an ephemeral existence, since molecular motion would tend to destroy it rather than create it: the liquid state is, after all, stable at this temperature. However, when the temperature falls slightly below T_f, the transient embryos have a better chance to grow sufficiently to become a stable seed and the crystallization of the liquid begins. If seeds do not form for any reason, the liquid remains 'supercooled' below T_f, but this state is very unstable, except in unusual cases, and is only observed when the temperature is just below T_f, for the probability of formation of a seed increases quickly as the temperature falls around this point. Note also that, as soon as crystals are formed, the heat released by the crystallization returns the temperature to T_f and we are back to the simple process of the transformation of liquid to crystal at constant temperature[1]. When the temperature falls further, the probability of formation of a seed, after having increased, passes through a maximum and then decreases because of the reduced thermal motion.

In certain cases, however, it is possible to preserve a supercooled liquid very much below T_f: in other words, to bring it down to a temperature where the formation of seeds is very improbable. In a normal liquid, this can only be achieved under special conditions. For instance, water dispersed into very fine droplets (several μm diameter) remains liquid down to $-40\ °C$. There is very little chance of a seed forming in the small volume of the droplets and, even if it does, its effect is strictly limited because it can only crystallize that particular droplet: a relatively negligible volume.

However, there are several liquids in which it is not only probable that crystallization will *not* occur but in fact quite normal. This happens with some of the very viscous liquids such as glycerol, for example. A crystal of glycerol *melts* at 5 °C but, when *cooled* without special precautions, liquid glycerol remains fluid well below 5 °C: it simply becomes more and more viscous, and eventually quite firm.

Fused silica is obtained by melting quartz at 1700 °C. At this temperature, its coefficient of viscosity is 10^8 times greater than that of water, such an enormous value being due to the nature of the interatomic bonds. Crystalline silica consists of SiO_4 tetrahedra, linked by strong covalent O–Si–O bonds (p. 100). In the liquid, some of these bonds are broken and others distorted but it does nevertheless preserve some rigidity. The crystalline embryos then have great difficulty in growing and, as the silica cools, it passes through T_f without the appearance of seeds until it reaches so low a temperature that the probability of their formation is negligible: *there is no crystallization*. However, the viscosity increases as the temperature falls and eventually reaches such a value that the material behaves like a solid since, in practice, it does not flow. Silica glass is a typical example of an *amorphous solid*.

The structure of amorphous solids

There is relatively little to say on this subject that we do not already know, because from the way they are formed we can deduce that *the arrangement of atoms in them is effectively the same as in the corresponding liquid.*

This is verified by X-ray diffraction as Fig. 8.2 shows. The average structures are defined by a distribution function (p. 139), and the values for the same substance in the liquid and amorphous solid states show only minor differences. As we have already

[1] This qualitative description must not be taken for an explanation of the *sharp transition* at T_f in terms of structural models. As we have already said, there is no known simple explanation of it.

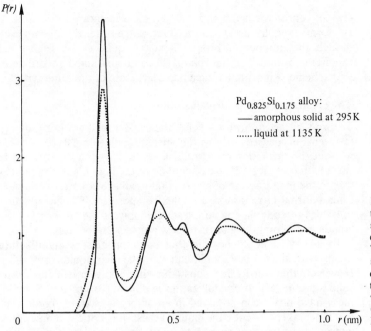

Pd$_{0.825}$Si$_{0.175}$ alloy:

—— amorphous solid at 295 K

...... liquid at 1135 K

Fig. 8.2 Radial distribution function for atoms in an amorphous solid and a liquid of the same composition, an alloy Pd$_{0.825}$Si$_{0.175}$. The average structures of the two substances are very close to each other, the short-range order in the liquid being slightly less well developed than in the solid. (P. Andonov, Laboratory of the Physics of Solids, Orsay.)

mentioned in the case of liquids, X-ray diffraction does not give a direct image of the structure. Certainly, if there happens to be only one type of atom present, the patterns of Fig. 8.2 will yield a statistical distribution of the distances between pairs of atoms. However, the great majority of amorphous solids have a complex composition with several types of atom, and it is clear that the different types will not follow the same statistical laws because of the differences in their diameters and types of bond. Unfortunately, it is not possible to deduce these separately from the data in a single X-ray pattern. The results show some improvement if the X-ray diffraction patterns are combined with those using neutron diffraction. In general, however, such techniques cannot give an accurate and certain picture of the average environment around an atom in any amorphous solid we care to choose. All we can do, therefore, is to postulate a model and adjust it so that it fits the observations as closely as possible. The trouble with this approach is that, even if perfect agreement is reached, the model is still only a *possible* one and others may show just as good agreement with experiment.

When dealing with liquids, we concentrated very much on those with close-packed disordered structures which represent the great majority of them, whether they are metallic or molecular. This type of structure is observed in amorphous solids as well, but some do possess covalent bonds which are *directed* and prevent the structure from being close-packed. Silicon, for example, crystallizes like diamond (p. 6): an atom is bound to four neighbours situated at the corners of a regular tetrahedron. These basic units still exist in the currently accepted structure of amorphous silicon, although with slight distortions (a few degrees fluctuation in the angles and a few per cent in the lengths). However, the elementary tetrahedra do not form a regular array, so that long-range order disappears even though the number of broken bonds is quite small. Figure 8.3 represents a model of amorphous silicon constructed according to that scheme. Amorphous silica has a structure of the same type.

The creation of new amorphous solids

To produce an amorphous solid, the critical temperature range (a few tens of degrees below the melting point) must be crossed without the formation of crystalline seeds. This can be achieved in silica with quite a slow rate of cooling of the liquid, and the same is true in the manufacture of glasses that are more common than pure silica but that have a more complex composition. The introduction of oxides of sodium, calcium, lead, etc., lowers the viscosity of the liquids and allows the glasses to flow at lower temperatures.

In a way, it could be said that common glasses manufacture themselves, since it is very difficult to make their liquids crystallize. However, there are other substances, normally crystalline in the solid state, in which crystallization is difficult to *prevent*. Some of these have now been prepared in an amorphous state. To do this, the time taken to cool the liquid across the critical temperature interval must be short enough to avoid the formation of crystalline seeds.

For silica or the common glasses, it is sufficient to allow the liquid to cool naturally in the atmosphere, although even here, if the

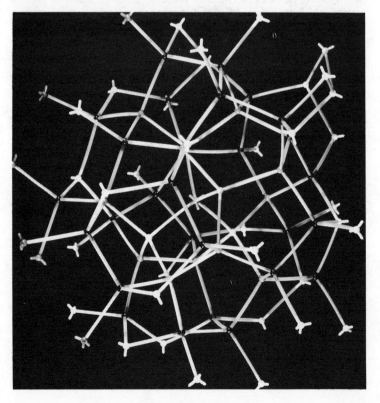

Fig. 8.3 Representation of the structure of amorphous silicon. Each atom is surrounded by four others at the corners of a tetrahedron, but one that is not quite regular. The whole structure is disordered in spite of a certain degree of order among the nearest neighbours. (R. Mosseri, National Centre for Scientific Research (CNRS), Bellevue, France.)

(a)

(b)

|← 1 cm →|

cooling rate is exceptionally slow, crystals can develop inside the vitreous mass (Fig. 8.4). For other liquids, *quenching* is essential: they are put abruptly into contact with a cold environment in such a way that their heat can disperse very rapidly. One such method squashes a droplet of molten metal (diameter 1 mm) between two copper plates at room temperature, achieving a cooling rate of 1000 °C per millisecond. Another method, illustrated in Fig. 8.5, is to allow a fine stream of molten metal to fall on to a copper wheel which is rotating at high speed. The filament of liquid solidifies into a ribbon which is given such a velocity that it detaches itself from the wheel and is flung off as shown.

New materials such as amorphous metallic alloys have been produced by similar methods using *ultra-rapid quenching*. Such alloys, which started as laboratory curiosities when discovered by P. Duwez in 1960, are now beginning to be manufactured industrially because they have some useful special properties and are cheap to produce. Ribbons of various compositions are available commercially, although details of the manufacturing process have not been published.

Other methods exist for obtaining amorphous substances, for instance by starting with a gas rather than a liquid. A vapour is first produced by heating a crystalline solid or a liquid, or by bombarding a solid surface with ions to detach atoms from it (this is 'cathode sputtering', which occurs at the electrodes in high voltage gas discharges in a low pressure enclosure). If the vapour,

Fig. 8.4 (a) A piece of glass that has been cooled exceptionally slowly by accident. It is partially crystallized in the opaque regions, which contain numerous crystallites surrounded by the clear regions of the amorphous glass. (b) Photograph of a small portion of 'Coupe du Grand Dauphin' (17th century, Museum of the Louvre). It shows the glass partially recrystallized for some unknown reason. (C. Lahanier, Research Laboratory of the Museums of France.)

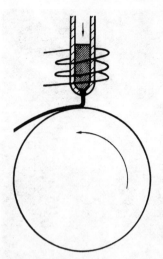

Fig. 8.5 Apparatus for producing an amorphous metal by ultra-rapid quenching. The metal is heated in a quartz tube by induction and is made to flow out of the bottom by gaseous pressure applied at the upper surface. Solidification takes place on a copper drum rotating at high speed. (Bastick and Tète, Laboratory of the Physics of Solids, School of Mines, Nancy.)

however produced, is made to come into contact with a cold surface, any atoms colliding with it remain stuck there. The low temperature reduces thermal motion so that there is no tendency for the atoms, which are deposited at random and are therefore disordered, to take up a regular arrangement. An amorphous layer is thus obtained, although only very thin films of thickness less than 1 μm can be prepared in this way. Nevertheless, it is a more effective method than quenching a liquid and it can succeed in 'amorphizing' some substances where quenching fails, however great the speed of cooling. This is the case for germanium.

A further process to consider is that of precipitation of a solid as a result of a chemical reaction between liquids. In general, the precipitate is a crystalline powder, but it can be amorphous if the atoms have a strong affinity for each other and stick together without taking up a regular structure. This is often a disadvantage for the chemist because the compound is much more difficult to characterize than a crystal, but the product may be interesting just because of its amorphous state. One example is that of a solution of sodium silicate treated with an acid to give a precipitate of amorphous hydrated silica. This, after dehydration, yields the 'silica gel' used as a drying agent. Another example is the reduction of a nickel salt by sodium hypophosphite which deposits a hard and shiny precipitate on objects immersed in the solution: a layer, not of pure nickel, but of an amorphous nickel–phosphorus alloy. This is a method used in industry because of its advantages over ordinary nickel-plating.

There are some substances that cannot be 'amorphized', no matter what procedure is adopted. The great majority of solids exist only in the crystalline state, so that the amorphous state is exceptional and peculiar to certain materials. Thus, while it is possible to obtain certain complex metallic alloys in durable amorphous form, it is out of the question with pure metals, save for rare exceptions. One of these is molybdenum, which can be prepared in the form of a thin amorphous layer by deposition from the vapour on to a surface cooled to liquid helium temperatures (below 4 K), but even here the layer recrystallizes as soon as the temperature rises. In the case of iron, there is a special chemical reaction which will produce a fine amorphous powder, but it is so unstable that it bursts spontaneously into flames in the atmosphere. It is completely impossible to obtain a piece of non-crystalline iron.

Why do metals crystallize so easily? It is because the atoms in the liquid state, although disordered, are stacked in a close-packed arrangement not too dissimilar to the crystalline form; the atoms are also mobile, so that the formation of the initial embryos takes place very frequently. Moreover, the growth to a stable seed is possible because it relieves the strains around the defects of the disordered liquid state. As soon as the temperature falls below T_f, there is nothing to prevent the metal from crystallizing, except perhaps when the atoms condense at temperatures near absolute zero.

In general, it is the same with compounds having purely ionic bonds (NaCl type). These also do not exist in amorphous form, in this case because the need for local electrical neutrality entails the alternation of +ve and −ve ions in the liquid. This greatly facilitates the formation of embryos having the regular crystalline arrangement.

Amorphous and microcrystalline structures

We have identified amorphous solids simply by the fact that they are not crystalline: in other words, their X-ray diffraction patterns do not consist of a series of sharp lines, which would be characteristic of a powder containing small crystals (Fig. 8.6). The diffraction pattern of an amorphous solid, like that of a liquid, consists of several broad bands which become wider as the diffraction angle increases. It would seem that this should provide a clear experimental way of distinguishing the two states and that there could be no ambiguity.

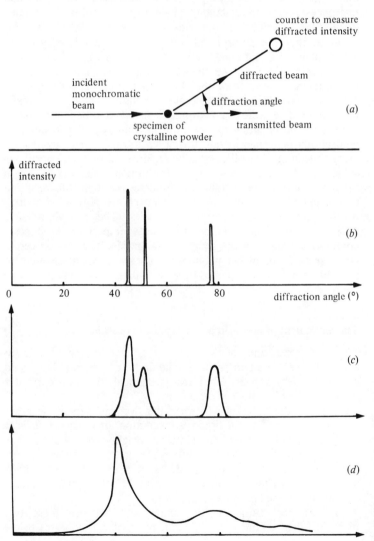

Fig. 8.6 Comparison of X-ray diffraction patterns of a substance in the form of small crystals with that of the same substance in an amorphous solid state. (a) Principle of obtaining the diffraction pattern of a microcrystalline solid. (b) Diffraction pattern from a crystalline powder with crystals of about 1 μm across. (c) The same with crystals 5 nm across. (d) The same solid in an amorphous state.

However, as the size of the crystals in the powder decreases, the width of the diffraction lines increases in inverse ratio to their diameters. The sort of pattern that would be given by a collection of crystal embryos, each containing no more than a few dozen atoms, can be calculated and is found to be quite close to that yielded by an amorphous solid in practice. However, even the *idea* of a crystal does not make a great deal of sense for such small sizes: how can the long-range order characteristic of a crystal be defined in such small regions, particularly when the atoms on the outside of them are bound to be greatly perturbed? Besides, the accuracy with which diffraction patterns can be measured nowadays enables us to assert that, whatever the size and shape of the small crystals, the pattern yielded by a crystalline powder can never be made identical with that from an amorphous solid.

However, before advances in techniques had allowed us to draw that conclusion, the similarity between patterns of amorphous and microcrystalline materials had given rise to confusion between them. One example of that was the assertion, for glasses for instance, that the amorphous state was only a microcrystalline one. Yet an essential characteristic of the structure of glass is its *homogeneity*, whereas the theory of microcrystals distinguishes the ordered regions from the disordered ones between crystals. Another example was that some thin films, which were really microcrystalline, were often wrongly considered to be amorphous.

Diffraction phenomena are not the only way to distinguish amorphous materials having their own intrinsic structure from a mere collection of small crystalline regions. By heating amorphous substances to certain temperatures for certain times that vary with the material, they recrystallize: crystal seeds are first formed and then rapid growth occurs until the amorphous material is entirely transformed into a polycrystalline solid. The behaviour of a microcrystalline solid is different: a rise in temperature produces a continuous growth of the small crystal grains. The two processes give rise to different behaviour when the materials are heated, so that they can be distinguished by what is called differential thermal analysis.

The movement of atoms in an amorphous substance

The relative positions of the atoms at a given instant of time (to about 10^{-13} s!) in an amorphous solid and the corresponding liquid are in fact very similar, often nearly identical. What differentiates the two is the *movement* of the atoms.

To make the difference between the two cases quite clear, we consider extremes. In the liquid, a given atom or molecule *diffuses* by virtue of its movement and that of its neighbours—that is, it has an irregular path that takes it further from its starting point as time goes on. In an amorphous solid, as in a crystal, the atom or molecule has a *site*—it vibrates continuously but remains 'encaged' in a small volume around a fixed point.

In fact, real situations are some way between these two over-simplified pictures. In a liquid, the molecule vibrates in the 'cage' formed by its neighbours for a certain time τ before jumping to a neighbouring cage: this is the basic diffusion process. In an amorphous solid, the time τ becomes very long but a jump into a neighbouring site is nevertheless possible from time to time. The characteristic time τ varies with temperature in a given substance and, at a given temperature, depends on the nature of the material. These are the dynamic aspects of the structure of amorphous substances which account for the variation of their properties with temperature.

The glass transition temperature
The basic idea in our models of amorphous solid structures has been the *continuity* between the liquid and solid states. Given that essential feature, however, we must now examine more carefully the experimental data on the variation of certain physical properties as a fall in temperature causes the change from liquid to solid. It is true that there is no discontinuity, but *the rate of variation of a property with temperature is not uniform.*

Figure 8.7, for example, shows the variations in the volume per unit mass (the inverse of the density), the coefficient of expansion and the specific heat capacity for the same substance, glucose, in both the amorphous and crystalline forms, the latter showing the discontinuity that we know occurs at the melting point (415 K). It is clear that in the amorphous form both the thermal expansion and the heat capacity show an abrupt variation over a narrow range of temperatures, while changing quite slowly outside it. For both properties, the critical region is centred around the same temperature, which is thus characteristic of the substance and is called the *glass transition temperature, T_g.*

Such a transition occurs generally in the amorphous state. It does not correspond to any change in the atomic structure: we have already noted that X-ray diffraction patterns are much the same for liquid and amorphous solid, and they certainly show nothing out of the ordinary in passing through T_g.

What does change from one side of T_g to the other is the nature of the molecular movements. Below T_g, the characteristic time τ of the molecule is large, while above T_g it is small—but large or small with respect to what? As our standard for comparison, let us use a time which is characteristic of the scale of the phenomenon and which is necessarily but implicitly involved in the measurement of the property under consideration. A value of 0.1 s has been proposed as an order of magnitude but 1 s could be used just as well since the definition is quite imprecise. It must not be forgotten, however, that τ increases 10^9 times when passing from liquid to solid and also that the temperature T_g is not exactly defined.

In the temperature interval between T_g and the melting point of the crystal, T_f, the material could be said to be in a supercooled liquid state and is so viscous that it *appears* solid, whereas below T_g, the material *is* solid. Indeed, such a situation is precisely that shown in Fig. 8.7 in the cases of the coefficient of expansion and the specific heat capacity: below T_g, the values are close to those of the crystal; above T_g, to those of the liquid. However, this simple

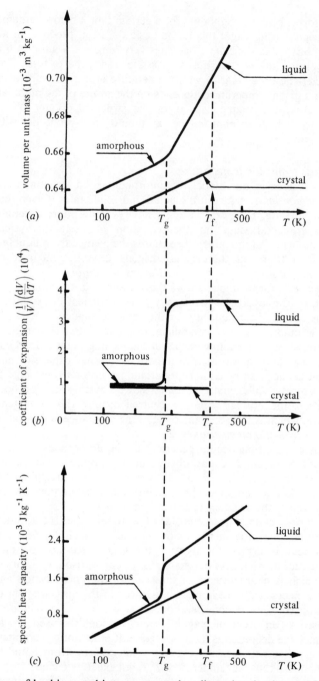

Fig. 8.7 Defining the glass transition temperature, T_g. Variations in: (a) the volume per unit mass, (b) the coefficient of expansion and (c) the specific heat capacity for glucose, which exists in both the amorphous and crystalline states.

way of looking at things must not be allowed to lead to confusion: the dynamic structure certainly changes, but always by a continuous evolution and never by a sharp transition.

Metastability in amorphous solids

Below the melting point, the only stable state of a substance is the crystalline one, so it cannot be stable if it is maintained in the form of a supercooled liquid: in fact, it can crystallize spontaneously, and in general, if special precautions are not taken, this occurs quite rapidly on our normal time scale.

In the same way, if we have managed to produce an amorphous solid from a liquid, it is no longer in a stable state and may also crystallize spontaneously, so returning to the form with greatest stability. However, if the material is kept at a low temperature, the smaller atomic movements lower the chances of the formation and growth of the seeds necessary for crystallization, particularly in cases where the amorphous state occurs more readily. For instance, on our normal time scale of months or years, we know very well that ordinary glass shows no apparent modification at ordinary temperatures or even when heated. Glass *appears to be stable* and is said to be *metastable*.

Devitrification is sometimes observed. This is a total or partial transformation of a material from glass to small crystallites without the object breaking or changing its external shape. Internally, however, the juxtaposition of many small crystals with random orientations makes the material opaque. This can be observed in tubes of vitreous silica which have been heated above 500 °C.

In some cases, devitrification of glass can be deliberately induced by the addition of an impurity that encourages the formation of seeds. The products obtained, called 'pyrocerams' or 'vitrocerams', are interesting because their mixed amorphous and crystalline structure gives them a greater mechanical strength and a very small thermal expansion. Thus, they do not break, even if subjected to thermal shocks by very quick and uneven heating. Sometimes, too, the crystalline part consists of such fine grains that transmitted light is not scattered and the glass remains transparent.

When the amorphous solid is difficult to produce, as with metallic alloys, crystallization occurs very readily (virtually instantaneously above about 100 °C to 300 °C). Some substances (such as those of liquid nitrogen or even of liquid helium) can only be kept in an amorphous state at very low temperatures.

The transition from the disordered amorphous state to the crystalline one is inevitably discontinuous, with a sharp change in volume (small, but real enough) and a release of latent heat as in the crystallization of a liquid. The structures of liquids and amorphous solids are very similar, so that we should expect the transformation to the same final state (the crystal) to be accompanied by very similar quantities of latent heat, and this is in fact observed.

The importance of amorphous solids

As we have already pointed out, solids in an amorphous state are exceptions: they are metastable, and most of those we are familiar with have been artificially created. Nature offers a few examples, such as obsidian (a natural glass) and pumice stone, both of volcanic origin[1], and there are numerous natural amorphous substances in living matter to which we shall return in Chapter 9.

[1] Volcanic activity involves precisely the sudden cooling or quenching of molten material that is most likely to produce amorphous solids or glasses.

However, the commonest ones are manufactured, and the easiest ones to make—the glasses—have been in use for thousands of years. Recently, however, more and more new types of glass have been created and we naturally ask what the important properties of these materials are that make them so useful.

In the first place, glasses, because they pass continuously from liquid to solid as they cool, are very malleable but still firm at intermediate temperatures, so they can be shaped very easily. Many of the techniques that are used would be impossible if the material passed sharply from the completely fluid state to a hard solid. The art of the glass-worker rests entirely on the skilful choice of temperature to give the glass whatever malleability is needed. Metals are ductile, too (that is, they deform without breaking), but they need much greater forces to shape them than does hot glass. Metals are worked in powerful machines such as presses and rolling mills, whereas all that is needed to shape a red-hot mass of glass is the breath of a human (Fig. 8.8).

In the second place, an amorphous solid is *homogeneous* down to a scale of one nanometre or so. It is true that the disorder means that the surroundings of one atom differ a little from those of its near neighbours, but, if we consider adjacent volumes of, say, about 0.05 nm^3, no distinctive variations occur either in the density of atoms or in the structure. In complete contrast, a solid that consists of a collection of small crystals contains the boundaries between them, of disordered material, separating two regions that are well ordered but have different orientations. In this case, long-range order has also disappeared, but this is because of sharp jumps from one crystal to another and not, as in amorphous materials, because of *continuous* changes from one atom to its neighbour.

The homogeneity of an amorphous solid is on a small enough scale for it to behave like a *perfectly continuous and uniform medium* for a beam of light with a wavelength of the order of several hundred nanometres. Such a beam can therefore be propagated without being scattered and, if the absorption is sufficiently low, the glass will be transparent. The absorption could be appreciable at certain wavelengths only, in which case the glass is still transparent but coloured. Transparency of glass is, of course, the main reason why it is so useful and we see now that it is a property linked directly to its structure.

Suppose we want to make a pane of glass for a window out of silica. This is possible because the absorption of light over the whole spectrum is very low. However, the material must also be homogeneous, and for this there are two possibilities. One is to take a very perfect single crystal of quartz, slice off a thin sheet and polish it. The pane obtained in this way is perfectly transparent because such a good crystal will be very homogeneous, but its size will be limited by that of the quartz crystal (at most a few decimetres) and . . . the process is ruinously expensive.

The other possibility is to use a glass of amorphous silica which can easily be obtained in large sheets of uniform thickness with very smooth surfaces if produced at the right temperature, and

Fig. 8.8   A glass-blower at work.

whose cost of manufacture is very low. Our pane of amorphous glass can very cheaply replace the prohibitively expensive sheet of quartz. (By way of contrast, a sheet also made of pure silica, but in the form of small quartz crystals like a mass of white sand, is completely opaque.)

The same problem, whose solution seems obvious, occurs in another industrial process: that of the manufacture of *silicon* for solar cells that convert the sun's energy into electrical energy. The condition to be satisfied by the silicon here is the same: it must be perfectly homogeneous. At the moment, however, the process adopted is the expensive one of preparing large silicon crystals with great difficulty and cutting off thin slices with a considerable loss of material. This long and costly process has to be used because we know that platelets of polycrystalline silicon will not work: they have a heterogeneous texture that causes a deterioration in electronic properties. Recently, however, it has been found that thin slices of amorphous silicon give interesting results and the manufacture of this type of material having large surface areas is simple and economic. If the electronic properties can be improved, solar cells could shortly become very cheap and have an enormous influence on the utilization of solar energy[1].

[1] This is a very brief explanation. In fact, 'amorphous silicon' contains hydrogen, which plays a decisive role in its properties.

Turning to similar examples in other fields, the current development of 'Xerox' photocopying machines relies on the properties of *amorphous selenium* which, again, is easily produced. We can also mention once more the case of amorphous alloys which have only recently appeared on the scene but which have promising properties: amorphous steel wires are already commercially available having greater strength than the corresponding crystalline variety used at the moment. It is also quite possible that amorphous alloys could have important magnetic properties because they are isotropic, whereas normal crystalline metals are anisotropic.

All these examples show that amorphous substances have only recently begun to play an important role in the research into materials that are opening the way to new technologies.

9 Between order and disorder: polymers, liquid crystals and soaps

The basic idea that has most frequently recurred during our exploration of the structure of matter so far has been the fundamental distinction between states that are ordered and those that are disordered, the two being separated by a discontinuity which is impossible to avoid when passing from one to the other.

It is true that some modifications have had to be made to such a simple scheme. The ordered state of solids is never *quite* perfect, and we have even seen that the *slight* disorder caused by defects is responsible for some of the most important practical uses of crystals. At the other extreme, condensed states of matter cannot be *completely* disordered because of their high density: it is true that *long-range order* may have disappeared totally, but the short-range order that always exists between closely-packed molecules still remains.

Nevertheless, in spite of the occurrence of a little disorder in ordered states and of some order in disordered states, the distinction between the two remains clear-cut, and many substances can be immediately allocated to one or other of the two categories. Alongside these, however, are other materials in which a quite genuine **intermediate state of order** is indicated by research into their structure. They fall into two categories. Either they are not all that far from being crystalline, but contain large deviations from perfect order involving a high proportion of the atoms, or they have disordered structures which are far from being completely random but show some degree, if not of long-range order, then at least of medium-range order.

Such substances pose problems for our tidy classification of structures, but we cannot simply ignore them as laboratory curiosities since they form part of so many objects in daily use and occur as some of the main constituents of plants and animals, and even of ourselves.

It is easy enough to foresee that atomic structures that are neither completely regular nor completely irregular would be difficult to describe. Our experimental techniques are undoubtedly not adequate to determine with sufficient certainty the large number of parameters needed to describe such structures. What precise knowledge we *do* have is in any case quite recent, dating only from the last few decades, and is also still incomplete, for research into these materials is currently very active. This is because they are of considerable significance both to fundamental science and to everyday life, where they form part of so many important products.

The sorts of substances we are talking about are *plastics*, *rubbers*, *textiles*, and so on, many of which are artificially created: new ones are appearing all the time, causing new technologies to flourish such as those involving liquid crystals, whose practical applications only began some fifteen years ago. It is worth noting that the plastics industry, which only began in the thirties, now has a world-wide production with a total value greater than that of steel.

Polymers

The elementary 'particles' of the structures we have studied until now, whether ordered or disordered, have been either single atoms or molecules formed from a small number of atoms, not more than a few dozen on average. These particles have been bound to each other in a precisely defined pattern which X-ray diffraction has shown to exist even when the number of atoms in the molecule is very large, as in the proteins where it exceeds a thousand.

In complete contrast, the characteristic common to all the substances we are going to study in this section, whatever their properties, is the gigantic size of their molecules. Each of these, called a **macromolecule**, is in fact a *polymer chain*, formed by the linear repetition of a group of atoms called a *monomer*. Figure 9.1 gives several examples. The number of monomer groups forming the chain can be huge (more than 10 000) but it has no fixed value for a given species and can vary considerably depending on the conditions under which the polymer is prepared. Not only that, but there is another important feature of a macromolecule: *it has no fixed configuration* but can take on an almost infinite variety of shapes.

The polyethylene chain

We take as our first example the simplest macromolecule, *polyethylene*, which consists of a chain of CH_2 groups terminated at both ends by CH_3. The chemical bonds are obviously the same no matter how many CH_2 groups there are: the number can vary widely but in the commonest materials is of the order of 25 000.

Along the chain, the carbon atoms are linked to each other by covalent bonds as in a diamond crystal: the C—C distance is 0.154 nm and the angle between two bonds such as C_1—C_2 and C_2—C_3 is 109.5°, as in the carbon tetrahedron on p. 6 (Fig. 9.2). Two C—H bonds also emerge from each carbon, for instance from C_2, in the directions of the other corners of the tetrahedron. Now consider the third link in the chain: C_3—C_4 of Fig. 9.2(b). It also makes an angle of 109.5° with C_2—C_3, but there is no reason why C_4 itself should stay in the plane defined by $C_1C_2C_3$ since the only condition to be satisfied is the fixed angle between C_2—C_3 and C_3—C_4. Thus, C_3—C_4 can be any one of the generators of a cone with axis C_2—C_3 and a semi-vertical angle of $180° - 109.5° = 70.5°$. In the same way, the next link can take any orientation around C_3—C_4 as long as its angle with C_3—C_4 is 109.5°, and the process is repeated all along the chain. Even with only four links, the shapes that become possible can easily be seen to be extremely varied. For several thousand CH_2 groups, the whole chain is like a

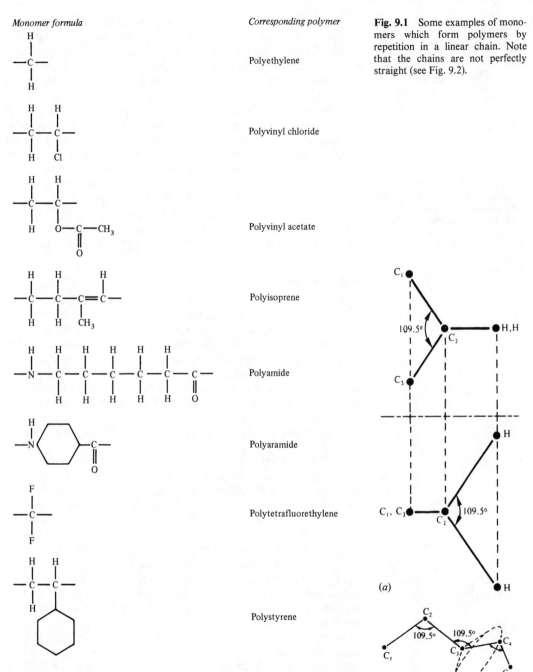

Monomer formula

Polyethylene

Polyvinyl chloride

Polyvinyl acetate

Polyisoprene

Polyamide

Polyaramide

Polytetrafluorethylene

Polystyrene

Corresponding polymer

Fig. 9.1 Some examples of mono-mers which form polymers by repetition in a linear chain. Note that the chains are not perfectly straight (see Fig. 9.2).

(a)

(b)

Fig. 9.2 (a) Plan view (above) and side view (below) of a CH_2 group in polyethylene and its bonds with neighbouring carbons. (b) Articulation of joints in a chain of five carbon atoms.

long thin thread that twists and turns in a very irregular fashion, from parts where it forms a tight ball to those where it is more or less perfectly straight (Fig. 9.3). The diameter of the chain, including the hydrogen atoms, is about 0.4 nm, and its length, for 25 000 carbon atoms, would be 3 μm if it were perfectly straight.

Since the angle between successive planes such as $C_1C_2C_3$ and $C_2C_3C_4$ is quite random, the chain is completely flexible, but the

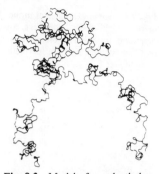

Fig. 9.3 Model of a polyethylene macromolecule containing 1000 CH_2 groups. The orientations of successive C—C bonds were chosen randomly as in Fig. 9.2(b). (L. R. G. Treloar and W. F. Archenhold, *Introduction to Polymer Science*, Wykeham.)

very randomness enables statistical calculations to be made. Distances between the extremities of the chain can be calculated for all the shapes it can take up and the statistical distribution of these distances can be ascertained. It is found that for a chain with n links each of length a, the most probable value of the distance is $a\sqrt{n}$.

This type of result recalls the statistical predictions made about the molecules of a perfect gas. Such predictions were possible because of the complete independence of the molecules and their resultant random motion. As soon as there was any interaction, however, the predictions became uncertain and complicated. It is the same with the polymeric chains: there are difficulties which stem from the fact that the linking bonds are not in practice oriented completely at random. It is quite obvious, for instance, that the chain cannot pass through a site already occupied by its own atoms (that is, it cannot pass through itself), so that some configurations are excluded.

Polymers in solution

In general, the molecules of a given polymer can be dispersed by certain liquids to form a solution. With very weak concentrations, the molecules are remote from each other, but as soon as the concentration exceeds 1% they begin to get tangled up with each other because they are so long. Moreover, the macromolecules, far from being in a vacuum, are in intimate contact with solvent molecules, whose interaction with them is far from negligible. For that reason, the chains may take up different shapes according to the nature of the solvent and they could be, on average, more tangled up or more stretched out than when not interacting with the solvent.

It is thus understandable that polymer solutions should have much more complicated properties than those of ordinary solutions in which the shape and size of the solute molecules remain fixed. One such property is *viscosity*. Even when very dilute, polymer solutions flow with great difficulty: this can be observed in solutions of gum arabic in water (a type of glue) or of rubber in toluene (the solution used for repairing inner tubes of tyres). Such high viscosity is explained in the following way: we have already seen (p. 146) that when a fluid, particularly a liquid, flows in contact with a solid wall, the various layers parallel to the wall do not move with the same speed. Viscosity is produced by the interaction forces between successive layers of molecules sliding over each other. However, macromolecules are so long that different parts of the same chain may be dragged along at different speeds, thus producing internal tensions in each molecule. Since the different parts may be in different liquid layers, this effect clearly links the various layers and greatly reduces the speed of flow. It is natural to assume from this picture that the viscosity must be proportional to the length of the polymer chains. In practice, the model is used the other way round: measurements of viscosity are made use of in estimating the chain length and thus the molecular mass of the polymer.

Crystalline polymers

When a pure polymer is in a condensed state (liquid or solid), the long chain molecules are tightly packed together and therefore tangled up with each other. This could lead us to the assumption that the structure of a solid polymer must be disordered. In fact, many of them can be considered as crystalline, having one essential feature similar to that of normal crystalline materials: the transformation to the liquid state occurs at a well-defined temperature and the transition is discontinuous. Even so, they cannot form a 'good' regular crystal because the elements that are stacked together are not identical: as we have seen, there is a variation in the size and configuration of the macromolecules.

Crystalline polymers have two specific features. Firstly, they always contain a *certain proportion of disordered domains* among the ordered ones. Secondly, *the ordered domains are not perfectly crystalline*. In the ordered regions, the surroundings of equivalent atoms are not strictly identical but show small variations. However, these are much less than in an amorphous substance, so that order is maintained over greater distances. As in a normal crystal, the basis of the structure is a unit cell with fixed atomic contents, but each cell in the whole array is slightly deformed in quite a random fashion, fluctuating about an average size and shape. This type of structure is sometimes called a 'paracrystal' (Fig. 9.4). X-ray diffraction patterns enable the average cell to be determined together with the mean positions of the atoms within it.

If the crystalline regions of the solid have their axes oriented haphazardly in any direction, the whole solid is isotropic and the polymer is said to be *non-oriented*. However, it is also possible for the crystals to have a texture with a more or less well-defined orientation and we shall see that these *oriented polymers* have particularly interesting properties. We now consider in turn examples of these two classes of crystalline polymers.

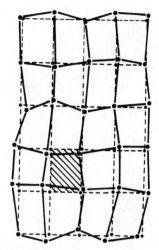

Fig. 9.4 Diagram of a 'paracrystal' in two dimensions, having one atom per unit cell. The smaller black dots give the ideal average atomic positions, and the larger ones the actual positions. The filled-in cell shows the mean shape and size about which the real cells fluctuate.

Non-oriented crystalline polymers

As a first example, we take *polyethylene* itself for a more detailed study. This has the simple chain already described above: $\ldots-CH_2-CH_2-CH_2-\ldots$. At ordinary temperatures, it is quite a soft solid, easily flexible and with a cloudy appearance. It melts at 129 °C, and the variation of its volume per unit mass with temperature (Fig. 9.5) shows the sharp discontinuity on melting. However, over an interval of some 10° below this there is an anomalous expansion of the solid as if it were undergoing a partial structural transformation preparatory to melting, a phenomenon that is not observed in normal crystals. Above 129 °C, polyethylene is a very viscous liquid, a state that is not unexpected since the tangling of the long chains must create internal friction as the liquid flows.

Solid polyethylene (commonly called 'polythene') is normally obtained by cooling from the liquid state, and its structure can be ascertained from its X-ray diffraction pattern. Comparison of the patterns in Fig. 9.6 shows that it is neither amorphous nor a normal

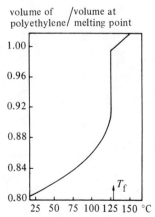

Fig. 9.5 Variation of the volume per unit mass of polyethylene with temperature.

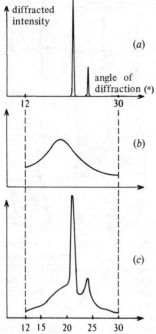

Fig. 9.6 (a) Part of the X-ray diffraction pattern of a well-crystallized powder of a saturated hydrocarbon C_nH_{2n+2} ($n = 12$). (b) Pattern of a completely amorphous substance. (c) Pattern of a partially crystalline polyethylene.

polycrystalline solid. The conclusions to be drawn from the pattern in Fig. 9.6(c) are precisely those already stated: part of the structure is disordered and part is ordered but with a significant lattice distortion. The pattern also shows that in the small ordered domains the polyethylene chains are parallel and are arranged in the same way as are the $-CH_2-$ chains in crystals of paraffins (C_nH_{2n+2}) having a small number of atoms in the molecule. The difference between them is that for $n < 20$ the crystals are well formed, whereas irregularities occur with the macromolecules. As happens in many organic materials, the paraffin chains are stacked in 'herring-bone' fashion (Fig. 9.7). It is also found that the crystalline regions are very small: in particular, their length along the direction of the chains (10–20 nm) is much less than that of the chains themselves (3 μm). Thus, each chain must traverse a large number of the crystalline regions and between two such regions must have a completely irregular shape. Finally, since the solid is isotropic, we know that the orientations of the chains in neighbouring crystalline regions must be different and, on average, quite random. From all these facts, we obtain the model shown in Fig. 9.8 for the structure of polyethylene.

The proportion of crystalline material in a solid polymer is called its *index of crystallinity*, and it can be determined by several methods, including the use of X-ray diffraction. The results are not exact, but cannot be so in any case, because in reality the structure is not simply an agglomeration of two distinct phases: there is a continuous increase in disorder from the centre of a crystalline region to its boundary.

In pure polyethylenes consisting only of $-CH_2-$ chains, the ordered regions can be quite extensive with an index of crystallinity as much as 90%. Some polyethylenes, however, have chains where some of the hydrogen atoms are replaced by a hydrocarbon or other radical, the simplest being $-CH_3$. These are called *branched polyethylenes*. The lateral branches disturb the regular stacking of parallel chains and so, where they become numerous, the degree of crystallinity decreases. If there is one lateral $-CH_3$ for every two carbons in the chain, and if they are randomly situated along the chain, the polymer is completely amorphous and is called *atactic* (Fig. 9.9(a)). However, if the $-CH_3$ branches are all in the same relative positions along the chain (Fig. 9.9(b)), the resultant uniform structure can yield crystalline regions (*isotactic* polymers).

Properties of polyethylene in relation to its structure
The *cloudy or milky appearance* of solid polyethylene is due to its heterogeneous structure with coexisting ordered and disordered domains. If it were completely amorphous, it would be transparent. Its *flexibility* is due to the presence of regions that are easily deformed because they are disordered. It should be noted that non-branched polyethylenes having very uniform chains and a high index of crystallinity (high density polyethylene) are harder, less flexible, and melt at higher temperatures than do low density materials with their non-uniform chains and smaller crystallinity.

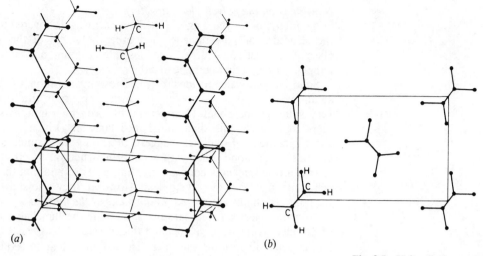

(a)   (b)

Fig. 9.7 Unit cell of a crystal of a paraffin C_nH_{2n+2} viewed: (a) at right-angles to the chains and (b) along the chains.

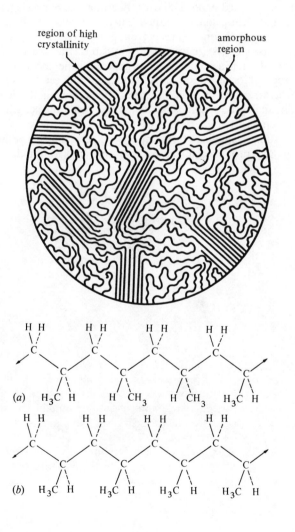

region of high crystallinity

amorphous region

Fig. 9.8 Diagram showing the structure of a partially crystalline polyethylene with well-ordered regions surrounded by amorphous ones. (H. W. Hayden, W. G. Moffat and J. Wulff, *The Structure and Properties of Materials*, Wiley.)

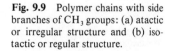

Fig. 9.9 Polymer chains with side branches of CH_3 groups: (a) atactic or irregular structure and (b) iso-tactic or regular structure.

New ideas on the polyethylene structure

Some fifteen or so years ago, something new was discovered about the structure of polyethylenes and other polymers of the same class. We must, of course, take this into account but, as often happens, it has rather complicated the problem.

Solid polyethylene formed by the slow evaporation of dilute solutions (instead of by cooling the melt) appears in the form of very small grains about 1 μm across. Electron micrographs show them to be very thin diamond-shaped lamellae (Fig. 9.10). The sharp geometrical outlines suggest that they are small single crystals, even 'good' ones with a uniform structure. Electron and X-ray diffraction show conclusively that this is so and that the unit cell is the usual one of polyethylene. However, they also show that the polymer chains are *normal to the plane of the lamellae*.

Now the electron micrographs show that the lamellae consist of parallel layers with well-defined thicknesses of between 10 and 20 nm depending on the specimen. The polymer chain is 100 times longer than this. To reconcile these two facts, the model proposed is of a chain folded on itself like a concertina, the parallel strands of equal length being joined at their ends by loops which can only consist of three or four $-CH_2-$ groups (Fig. 9.11). However, even though this idea is geometrically possible, the reason for the adoption of such a complicated structure by the substance is not understood. Nor is the value of about 10 nm for the thickness of the layers, since it is both large compared with the length of an individual group and small compared with the length of the polymer chain.

The same lamellae have not been certainly identified in polyethylene produced in the mass from a cooled liquid, but it is

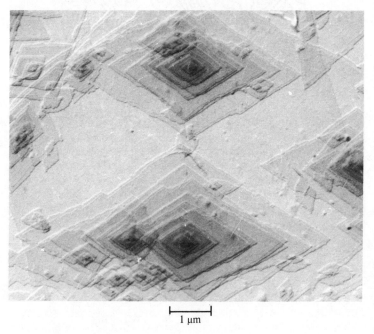

Fig. 9.10 Electron micrograph of polyethylene crystals obtained by crystallization from solution. They are remarkable for their geometrical shapes and their arrangement in parallel layers of uniform thickness of about 10 nm. (A. Keller, University of Bristol.)

1 μm

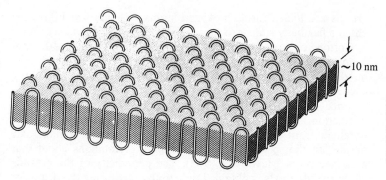

Fig. 9.11 The folding of chains in the crystalline polyethylene layers of Fig. 9.10. (H. W. Hayden, W. G. Moffat and J. Wulff, *The Structure and Properties of Materials*, Wiley.)

thought that the same type of arrangement could be involved although probably with less regularity. The folding of the chains would then occur in the ordered regions in the model of Fig. 9.8.

Another observation that has been made in polyethylene crystallized from the liquid is the existence of what are called 'spherulites': groupings of small crystals that develop radially from a central point over a distance of about one-hundredth of a mm. Such long-range order must have a significant effect on the properties of the polyethylene, but its origin has not yet been explained.

In all, then, solid polyethylene contains arrangements having types of order on very different scales all superposed on each other—the order between polymer chains, the folding of these chains, and the arrangement of the small crystalline domains—none of which are quite perfect. This gives some idea of the complexity of the problems to be solved, and of their importance too, for each of the elements of structure is a factor which determines the properties of the synthetic material.

Oriented crystalline polymers

In this class of polymer, the crystalline regions are still mixed up with amorphous ones but no longer have random orientation. The most important case, because of its applications, is that of a **fibre**, characterized by the fact that, on the whole, *the polymer chains have a common direction, the fibre axis*. In reality, this is only approximately true: firstly, because the chains can be inclined at a few degrees to the axis and, secondly, because a certain proportion of chains have random directions in the disordered regions. Materials of this sort are very well known and include the natural polymer fibres (cotton, wool, silk, etc.) and synthetic fibres (nylon, etc.). We shall first examine in detail a very typical example, that of a cotton fibre.

A natural fibre: cotton

Cotton is a natural fibre found in the stems or fruits of certain plants and consists of a chemical substance called *cellulose*. This is a chain polymer whose monomer is formed from two glucose nuclei bound to each other through an oxygen atom but with the elimination of a water molecule (Fig. 9.12): a complex structure with 42 atoms and a length of 1.03 nm. Two successive monomer

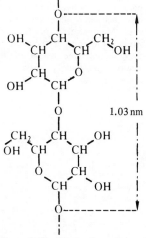

Fig. 9.12 The formula of cellulose.

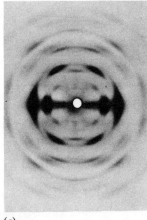

(a)

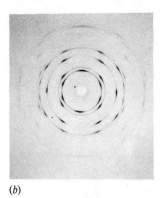

(b)

Fig. 9.13 (a) X-ray diffraction pattern of a cellulose fibre (China-grass). (b) X-ray diffraction pattern of a drawn aluminium wire. The spots on (a) are very similar to those on (b), where all the crystals have their axes along that of the wire. Measurement of the gap between the lines of spots on (a) gives directly the length of the cellulose molecule. (Institute of Textiles, Paris.)

groups are joined through oxygen atoms as joints which give the chain a flexibility similar to that of the simpler polyethylene chain.

Cotton occurs in the form of very fine fibres (<0.1 mm diameter) with variable lengths of the order of a centimetre, but bundles of these forming long thicker thread are more commonly encountered.

Once again, the atomic structure is revealed by using X-ray diffraction. The patterns obtained (Fig. 9.13) are similar to those obtained from collections of crystals having one common axis parallel to that of the fibre but taking all orientations about that axis (see fibrous texture, p. 133). The periodicity of the crystals along the fibre axis is obtained directly by measuring the distances between groups of spots on the patterns. The value found for this is 1.03 nm, precisely the length of the cellulose chain monomer. We are thus led to the following structure: the cellulose chains are drawn out along the fibre axis; they are arranged uniformly, as in a crystal; the small crystal domains have a common axis along the fibre but random orientations about that axis. A detailed study of the X-ray pattern also gives the lateral distances between chains and their relative positions: that is, it gives the crystalline unit cell and the positions of the atoms within that cell (Fig. 9.14).

The cotton fibres that we see with the naked eye are thus the outward appearance on our scale of the chain-like molecular structure, in exactly the same way that the plane faces of a crystalline mineral are the visual aspect of its geometrical lattice of atoms. The very finest fibrilla that can be detached from a fibre is still a collection of a very large number of very small crystals whose chains have a common axis: it is not possible to isolate a single crystal.

The structural model of cotton fibres as built up from small crystals is only a first approximation, as it was for non-oriented polymers. It is clear from Fig. 9.13 that there are significant differences between the pattern of a cellulose fibre and that of a well-crystallized metal such as aluminium which also has a fibrous texture. Using X-ray diffraction theory, it can be shown from the data provided by these patterns that *the cellulose crystal is highly imperfect*. In the first place, the individual crystals in the fibre are certainly *very small*: their length corresponds to a chain of at most 50 monomer groups and their width to less than five parallel chains. Moreover, even in the small crystalline regions the order within chains is not perfect: distances between equivalent atoms, which should be the same from one unit cell to the next, fluctuate slightly around a mean value.

Finally, *there are disordered regions between the ordered ones*. The length of a chain is considerably greater than that of a crystalline domain, so the chain *must* pass from one crystal to another as illustrated in the diagram of Fig. 9.8, with the difference that the direction of the chains in all the ordered regions is the same. The proportion of amorphous material in cellulose can be put at 30%, with the same reservations as already made in the case of non-oriented polymers.

Natural cotton fibres are flexible enough to be twisted into a thread of constant cross-section used for weaving. The length of the natural fibres varies according to the quality of the cotton. The longer they are, the more a thread will have a smooth surface because there are fewer joins between fibres in a given length. That is why a fabric made from long-fibred cotton appears soft and shiny, while one made from shorter-fibred cotton appears fluffy.

Synthetic fibres

The valuable and important properties of natural cotton fibre stem from the common orientation of its crystals, and so attempts have been made to reproduce such properties, and even improve on them, by synthesizing polymers with very good fibrous textures.

Rayon

The cellulose fibres in cotton have a limited length, and it would clearly be better to have continuous fibres of virtually infinite length available. This has been successfully achieved with *rayon*. The starting material is cellulose from wood pulp, which is very poorly crystalline and whose chains are not oriented. The cellulose is treated chemically and transformed into a highly viscous mass from which fine threads are drawn. Within these threads, cellulose is regenerated with chains along the axis. This was the first of the artificial textile fibres to be produced (1910) and it has since appeared in various forms such as artificial silk, viscose, and so on. The quality of the fibres is quite variable and depends on the degree of crystallinity and orientation that has been achieved: it is quite clear that the essential factor is the order and parallelism of the cellulose chains.

Oriented polyethylenes

There is a variety of techniques by which melted polyethylene can be made to solidify in the form of a continuous filament of uniform diameter. Such a filament has the same structure as a large solidified mass would have: that is, it is partially crystallized and without definite orientation. If, however, the filament is subjected to a continuously increasing tension, it suddenly stretches to two or three times its length at a certain value. At the same time, its cross-section diminishes in such a way that the volume remains sensibly constant. The transformation is certainly not continuous: at a tension below the critical value, the elongation is quite small (<10%), but as soon as the critical tension is reached, the polyethylene filament abruptly necks, i.e. becomes narrower, over a small section with a new diameter that is in fact its final value (Fig. 9.15). If the tension is then held constant, the narrowed part grows in length at the expense of the remainder of the original filament until all of it is drawn into the thinner form. At that stage, if the tension is increased still further, the additional elongation is quite small and is elastic. The drawn form of the filament has become very strong.

The drawing process induces an unfolding of polymer chains along the axis of the filament, thus forming small crystals with a good fibrous texture. Some regions still remain disordered, so that the structure is, on the whole, similar to that of cotton.

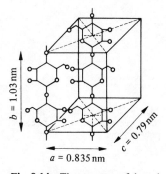

Fig. 9.14 The structure of the unit cell of cellulose, with the two molecules of Fig. 9.12 as motif. Oxygen atoms are represented by circles, carbon atoms are at the intersection of the lines representing bonds, while hydrogen atoms are omitted.

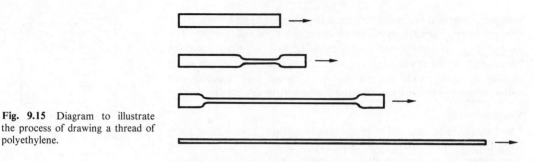

Fig. 9.15 Diagram to illustrate the process of drawing a thread of polyethylene.

Nylon

The plastics industry has developed a considerable number of different polymer formulae as well as a variety of production techniques, both of which have led to the present vast range of products on the market. Not only are artificial fibres easier to produce than natural ones, but in general many of their properties are superior.

One of the best known synthetic fibres is *nylon*. This is a polyamide whose monomer has the formula

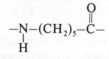

This chain is similar to that of polyethylene, but it is periodically interrupted by the $-CO-NH-$ groups, which are most important because they establish between the chains quite strong hydrogen bonds that do not exist in polyethylene. As in the latter material, however, the orientation of the crystalline chains is again produced by the same technique of stretching filaments.

The structure and properties of oriented polymers

However complex the polymer structures may be, some of their basic properties are easy to understand.

The cohesion between the ordered regions depends on the nature of the bonds between chains. In nylon, the hydrogen bonds, which form lateral bridges from one chain to another, account for the high melting point of 265 °C: that of polyethylene, with no such bonds, is only 129 °C. Cellulose also has hydrogen bonds: it does not soften on heating, nor does it melt, but it starts to decompose at about 200 °C. In addition, imperfections in the alignment of cellulose chains allow water to penetrate between them, thus giving cotton a practical property of enormous importance: *it can absorb water* which makes the fibres swell up but this is the property that makes cotton so pleasant to wear on the body. However, it dries slowly after being washed because the water is held between the cellulose chains by the chemical bonds.

A very important property of a fibre is its *tensile strength*. When a filament of oriented crystals is stretched, a large proportion of the chains directed along the axis is extended. This means that the opposition to the external tension is provided by the very strong bonds between carbon atoms, the same bonds that confer great

Table 9.1 Comparison of the mechanical strengths of metallic wires and polymer fibres. (A nylon thread of 1 mm² cross-section can support a person weighing 80 kg.)

Material	Breaking strength, R (10^7 N m^{-2})	Relative density, d	$\dfrac{R}{d}$
Steel piano wire	200	7.8	26
Aluminium wire	17	2.7	6.3
Rayon	30	1.56	19
Drawn nylon	100	1.15	87
Aramide fibre (high strength kevlar)	300	1.3	230

hardness on diamonds. However, there is one essential difference: in diamond, *all* the carbon atoms are linked to four neighbouring ones so that the crystal is strong whatever the direction of the applied forces. In the polymer, the only direction in which the bonds are effective is along the axis of the filament, and even in this direction some are ineffective because of the disorder in certain regions. Fibres with greater and greater strengths can be produced by improving the degree of crystallinity and orientation: a recent material called *kevlar* is an example, as Table 9.1 clearly demonstrates.

This table, in its first column, compares the breaking strengths of two metals (steel and aluminium) with those of three textile fibres (rayon, nylon and aramide) for filaments with the same cross-section, and shows that the strongest metal is only twice as good as nylon. However, if we compare filaments with the same mass per unit length rather than those with the same cross-section, the synthetic fibres have the advantage (last column of Table 9.1). Nylon is three times stronger than steel, and kevlar nine times stronger[1]. This last substance can be used to make bullet-proof jackets that are no heavier than ordinary clothes.

While the disordered regions of the fibres have the disadvantage that they lower the strength, they do have the advantage that they increase the flexibility, just as defects in perfectly crystalline metals increase their ductility. Those parts of the polymer chains crossing disordered regions are easily deformed. The quality of a textile thread, and of the cloth obtained after weaving, depends on the combination of tensile strength and flexibility of the polymer fibres.

Amorphous polymers

If crystallization is to be achieved, even if only partially, polymer chains need to be uniform over quite long lengths: in other words, there must be many identical repetitions of the basic monomer element. Any departure from this condition means that the chains cannot form compact uniform bundles. Two ways in which this can happen are, firstly, that the polymer chains consist of a random succession of several types of elementary monomer (producing what are called 'copolymers') and, secondly, that lateral branches

[1] We should also mention the case of *carbon fibres*, which are 6.5 times stronger than steel but are not polymers.

occur at various points along them. In neither case does the polymer crystallize to any great extent.

There are, however, other types of polymer whose X-ray diffraction patterns show an amorphous structure and we go on to discuss these now. First of all, we must be clear that these amorphous structures are quite different from those described in Chapter 8. In those cases, the structural element was an atom or a small molecule, while in a polymer it is a long chain built up from definite elements. Whatever the disorder *between* molecules, there exists, *at least along the chain*, an order of much greater range than that of the short-range order in condensed states with small molecules.

The structural model of an amorphous polymer is easy to picture in general terms, but is difficult to be precise about. Even if the chains are identical in composition, they take up random configurations and in the condensed state must be tangled up with each other. However, the most important properties are not determined only by the static features of the structure: it is so disordered in any case that it hardly varies with temperature or from one substance to another. What is important is the *relative motion of the chains* and the possibilities here depend on a balance between two influences: thermal motion, which clearly increases with temperature, and the intermolecular binding forces, which oppose the relative displacement of two chains.

As examples, we discuss two amorphous polymers whose properties at ordinary temperatures show a marked contrast with each other:

1. *Perspex* has properties similar to those of ordinary glass, i.e. it is hard, brittle, not easily deformed. The monomer of perspex is methyl methacrylate:

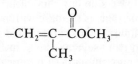

2. *Rubber* is soft, flexible and very elastic. Its formula is given later.

Both of these show marked changes if the temperature is altered: perspex heated to 140 °C becomes flexible and elastic, while rubber cooled to −100 °C becomes hard and brittle like glass. Amorphous polymers such as these cannot therefore be separated into two distinct classes. Instead, each of them passes from one state, the **vitreous** or **glassy** state, to another, the **rubber-like** or **rubbery** state, as the temperature rises. In the following sections we describe these states in more detail, but we note that there is in fact a transition temperature between them which is characteristic of each polymer. It is called the *glass transition temperature*, T_g, which we have already met in the last chapter in connection with simpler amorphous materials. As we indicated there, it is not the structure that changes at T_g but the character of the motion on an atomic scale. The transition is continuous, but is particularly rapid in a small temperature interval around T_g. In the same way that the glasses in Chapter 8 with small molecules were said to be in a particular state above T_g, that of a supercooled liquid of enormous viscosity, polymers above T_g are said to be in their rubbery state.

We must be a little more precise about the meaning of 'motion on the atomic scale' in polymers. At temperatures way *below* T_g, the general arrangement of each chain remains fixed in time, with the various monomer groups vibrating about a mean position. A long way *above* T_g, complete rotations of the monomer groups around the direction of a linking bond are possible, so that chains may alter their internal configuration as well as that relative to neighbouring chains. Around T_g, the polymer passes gradually from one regime to the other.

Glassy polymers
The existence of long-chain molecules in glassy polymers does not give them any particular properties different from those of ordinary glass. Hence, as long as the basic structural unit of the polymer does not absorb radiation in the visible spectrum, its perfectly homogeneous amorphous structure makes it transparent to light, as is glass. This is particularly the case with perspex, which has many applications based with good reason on its excellent transparency: the manufacture of spectacle lenses, for example. If the perspex were partially crystalline, it would be no more transparent than polyethylene.

Whereas glasses with small molecules can always crystallize, sometimes easily, sometimes with difficulty, many amorphous polymers do not exist in the crystalline state because the non-uniformity in the molecular chains completely prevents uniformity in packing.

However, some polymers do have sufficiently uniform chains for crystallization to occur. This is the case with natural rubber, which is amorphous at ordinary temperatures and which has a T_g of $-80\,°C$. When held for several days at $0\,°C$, it gradually crystallizes, but only partially and with very imperfect crystals: it loses its translucent quality and becomes hard and stiff. Crystal seeds are formed but their development is very slow because the motion of the chains is so limited. (In polyethylene, on the contrary, the generation and growth of seeds is so rapid that there is no amorphous state.) This crystalline rubber is stable at low temperatures but has no distinct melting point, its normal state being recovered at around $15\,°C$.

Rubber and its elasticity
Natural rubber, which is found in certain trees such as *Hevea brasiliensis*, is a chain polymer whose motif is

$$-CH_2-\underset{\underset{\displaystyle CH_3}{\displaystyle |}}{C}=CH-CH_2-$$

The differences between this and the polyethylene chain are, firstly, the presence of the CH_3 branch and, secondly, the presence of the double bond between every fourth pair of carbon atoms. Nowadays, synthetic rubbers of slightly different formulae are used, but they share with the natural product its most important and remarkable property, its *extraordinary* **elasticity**. Polymers of this sort are known generally as **elastomers**.

All solids are elastic: under tension, they stretch in proportion to any applied force and *recover the identical initial shape as soon as the external forces are removed*. This is true only up to a certain limiting force: one which produces only a small extension of the order of 1%. A piece of rubber, however, can *stretch perfectly elastically up to ten times its original length*. The disproportion between the two effects shows that their basic explanations must be completely different.

Consider a simple solid with its atoms in equilibrium positions separated by a distance a. When extended over its elastic region, the distances a will suffer an increase which can be measured macroscopically and is typically up to 1%. The atoms are then no longer in equilibrium positions and interatomic attraction will tend to restore them, thus opposing the external stresses. When these stresses are removed, therefore, the atoms return to their normal distance apart, a. It is clear, however, that the atoms could not be separated by very large distances compared with a without completely upsetting the structure, so elastic extension of such solids is always small: there is no question of a being extended to $10a$.

The mechanism which accounts for the elasticity of crystals is thus no good for explaining the elasticity of rubber. The origin of the latter lies in the existence of the long polymer chains. There is normally a lot of bending between any two points in a chain, so that when it is unfolded *the distance between them can increase considerably without bonds between neighbouring atoms being broken*. This is the basic process in the elastic extension of rubber.

However, this idea needs to be sharpened up. Firstly, there need to be points of attachment along the chains both to transmit the tensions and to prevent the rubber being fluid: what exactly *are* the points of attachment that play such a special role? Secondly, why is there reversibility between the stretched and unstretched states of the chains? Finally, why is it that *all* chain polymers are not elastic like rubber?

For the chains to be in such a state that they can unfold, their individual monomer groups must have great freedom of movement and so the temperature must be above the transition point T_g, as we have seen. If there is so much mobility and flexibility, why is rubber a solid and not a viscous liquid? In fact, natural rubber extracted from trees *is* quite close to a liquid state and it only acquires the unusual properties of a true rubber after the process of **vulcanization**.

As long ago as 1839, Goodyear found that natural rubber was considerably changed and improved by incorporating a small quantity of sulphur into it by heat treatment. This discovery, which started the rubber industry, was made purely empirically since nothing was known at that time about the atomic structure of rubber (or of other substances!). It is now known that the sulphur atoms establish very strong bridges between pairs of chains. If the sulphur concentration is high, about 30% for example, the segments of chain between the bridges become quite short and form such a dense network that the material becomes very hard: the rubber is transformed into a solid called *ebonite*.

However, if the sulphur concentration is low (0.5–3.0%), the bridges are quite a distance from each other and the segments of chain between them quite long enough to be deformable but not so long that they can slide over each other. The sulphur bridges form the 'points of attachment' that we were looking for (Fig. 9.16). When rubber is stretched, the network of attachment points elongates in the direction of the extension and the chains consequently unfold. This idea is geometrically very simple, but it does not explain why a force is needed to straighten the chains nor why they return spontaneously to their original curled-up configuration when the force is removed.

To understand these matters, we first analyse what is apparently a quite different phenomenon. Consider some gas molecules enclosed in a cylinder with a piston which is fixed in such a position that the gas pressure is equal to atmospheric pressure. The molecules inside the vessel are completely disordered: in particular, the position of each of them is completely unknown, except that it must be somewhere within the volume V (Fig. 9.17). It would therefore be *possible* for all of them to be in the lower half V_1 with the upper half V_2 completely empty. We know that this does not happen, and the explanation is provided by statistical physics: we can put it in the following rather complicated way. Any one 'configuration' of the gas specifies the positions of all the molecules, and all configurations have an equal chance of occurring because of the complete disorder. Now there are far more configurations possible in which the gas occupies the whole volume than there are with it occupying the half volume V_1. The first situation is thus vastly more *probable* than the second one. A general rule, which applies not only to the present problem but to all the phenomena observed in our world, is that *a system is in equilibrium when it is in its most probable state*. The gas will therefore occupy the whole volume V, its pressure balancing the atmospheric pressure on the piston. *Work must be done* to reduce the volume of the gas and make it occupy V_1, work that is provided by an external force. Conversely, if the piston starts in position V_1 so that the pressure is higher than atmospheric, then when released the gas expands spontaneously to a volume V.

Now let us return to rubber. The simple behaviour we are concerned with is that of a chain with n jointed groups between two points A and B (Fig. 9.18). For a given distance AB, the chain can theoretically take up a very large number of configurations and it can do this because of its great freedom of movement. The greater

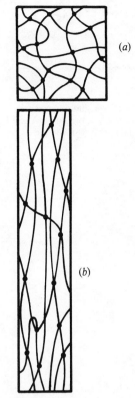

Fig. 9.16 Diagram to illustrate the structure of polymer chains in rubber linked by sulphur points: (a) normal state and (b) extended state.

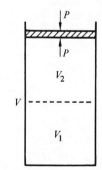

Fig. 9.17 Compression of a mass of gas.

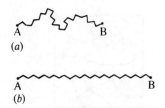

(a)

A   B
(b)

Fig. 9.18 A molecular chain of rubber between two points A and B: (a) slack and (b) completely extended.

the distance between A and B, the fewer will be the number of possible configurations: this is easy to see, because in the limit AB has a maximum value when the chain is as straight as it can be: in that case, there is only one possible configuration.

The shapes taken up by the molecular chains in rubber correspond to the configurations of the molecules in the gas. In both cases, configurations are changing all the time because of thermal motion. Increasing the distance between the ends of the flexible chain produces fewer probable states just as did decreasing the volume of the gas. *Neither of these operations will take place spontaneously: external work has to be done.* In compressing the gas, an external pressure is exerted on the piston which is itself being hit by molecules whose momentum is changed by the collisions. In stretching a molecular chain of rubber, the external tensile force is needed to change the momentum of the moving monomer groups so that they can pass from one configuration to another more elongated one. When the pressure on the gas is released, it expands spontaneously; in the same way, the chain contracts spontaneously when the tensile force is reduced.

The elasticity of rubber is not due to the binding forces between atoms as it is in a metal: it has a *kinetic origin*, as has the elasticity of a gas, and is connected with the disordered motion of the groups in the polymer chain just as it is with the disordered movements of the molecules in a gas. We can treat two such different systems by similar methods because of a very general statistical concept which we have only used qualitatively, almost as if it were a philosophical idea[1].

What is surprising about this is that physicists are able to draw quantitative conclusions from such ideas and compare them with experiment. Thus, just as a thermally isolated mass of gas heats up when sharply compressed and cools down when allowed to expand freely, so we predict that *a piece of rubber should heat up when stretched and cool when allowed to contract.* This is, in fact, what happens and, although known for a long time, it intrigued physicists for many years.

The theory also predicts to within 25% the coefficient of elasticity of rubber as a function of the number of bridging sulphur atoms without introducing a single empirical parameter. When we think of the complicated nature of the structure, and of the gross approximations and assumptions made in order to carry out the calculations, this particular success is reassuring as an indication that our knowledge of the structure of matter rests on secure foundations.

In its normal equilibrium state, rubber is completely amorphous, with its molecular chains curled-up and folded. When it is greatly extended, a significant proportion of the polymer chains are elongated in the direction of the stretching and become organized into small ordered domains. It could be said that rubber crystallizes

[1] Readers familiar with thermodynamics will recognize that the concept involved is the change in the entropy of a system at constant internal energy.

by being stretched. Of course, the crystallization is only partial, because there are still disordered regions and the crystals are imperfect, as they are in other fibrous polymers. X-ray diffraction patterns confirm this behaviour.

Thermosetting plastics

These polymers are produced by the condensation of two relatively simple molecules of different types which become attached end to end to form long chains. In addition, the chains may be branched with the lateral groups joined on to other chains. The whole structure is **reticulated**: that is, it forms an irregular network of chains, linked together by strong bridges and so interwoven that the material becomes very hard and difficult to deform even at high temperatures (Fig. 9.19).

The condensation is made to occur by heating since the monomer liquids that form the starting materials harden when the temperature is raised. This accounts for the description of the plastics as 'thermosetting' since they are unlike the non-reticulated amorphous polymers that soften on heating.

One product of this type, ebonite, has already been mentioned. In that material, the reticulation is accomplished by the sulphur atoms in the chains. There are now many thermosetting resins with different formulae: *bakelite*, for instance, is the result of a condensation of phenol with formaldehyde, and there is also urea-formaldehyde, melanin-formaldehyde, and so on. These were the materials from which so many of the first 'plastic' objects in everyday use were made. The great variety of properties possessed by plastics—their appearance, their strength, their hardness, and so on—is explained by the diversity of their constituents and of their methods of preparation.

In conclusion, we comment on the vast range of these organic polymers which we commonly call *plastics*. Not only can we create by chemical synthesis a virtually infinite number of products with different formulae, but we can also use various physical treatments to lead to quite diverse atomic structures (more crystalline or less, more oriented or less). From all this, a considerable number of new materials have been produced in the last thirty years or so, some of them currently holding a place in the forefront of modern technology.

It is interesting, in fact, to compare the immense field of what some have called 'plasturgy', with the relatively modest one of metallurgy. The metallurgist only handles about a dozen different types of atom, which are naturally abundant and not prohibitively expensive. While these can be combined to make alloys, these are all crystalline with only a small number of different atomic structures. There is not the slightest chance that a new pure metal will be discovered tomorrow, while the appearance of an alloy with new properties is becoming less frequent so that progress in this field is quite modest. It is worth noting that the discovery of amorphous alloys has been the most remarkable advance in recent years, and this is because their structure is so different from those of conventional alloys.

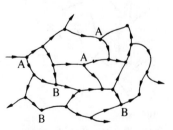

Fig. 9.19   Diagram to illustrate a reticulated polymer. The chains branch at points A and join up with other chains at points B.

Liquid crystals or ordered liquids

A little under one hundred years ago, something unusual was observed in a very straightforward experiment when a crystal of an organic substance called cholesteryl benzoate was melted. As expected, there was a discontinuity when the crystal was heated to 145.5 °C: it became a viscous and cloudy liquid. However, when the temperature was raised still further, a second discontinuity appeared, again at a definite temperature: the liquid became fluid and transparent at 178.5 °C. When cooled, the opposite transformations occurred at the same temperatures. At both temperatures, the changes were accompanied by an abrupt alteration in the volume and by absorption or emission of heat according to whether the temperature was rising or falling. The conclusion could thus be drawn that this substance showed *two* genuine phase changes in succession.

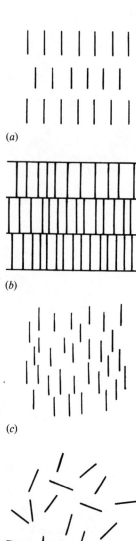

(a)

(b)

(c)

(d)

Fig. 9.20 Structures of the mesophases existing between crystal and liquid: (a) crystal, (b) smectic phase, (c) nematic phase and (d) isotropic liquid.

Intermediate states between crystal and liquid

Innumerable cases of melting had been studied until that time, but never before had such behaviour been observed. This was the first example showing that an intermediate state of matter between solid and liquid could exist. Although it is a relatively rare phenomenon, hundreds of organic substances are now known that possess an intermediate phase or even a succession of distinct phases. These are called **mesophases** or *mesomorphic* phases, and the substances themselves are usually called *liquid crystals*.

All substances with a mesophase have a feature in common: their molecules have an elongated rod-like shape, generally a little flattened, with a typical length of around 2.5 nm and a cross-section of 0.6×0.4 nm[1]. Another very important feature is that the molecules are rigid, at least in their central portions, and do not have the articulated joints of the $-CH_2-$ paraffin chains.

Molecules in condensed states of matter are packed closely together, so that by observing the way elongated rice grains behave in two dimensions (Fig. 7.5) we can see intuitively that lath-like molecules would tend to align themselves parallel to each other. This is certainly what happens in liquid crystals, all the more so because there are specific intermolecular interactions that favour the tendency. As usual, thermal agitation opposes this and as a result the actual structure depends on the balance between the two effects. It is clearly a delicate balance, because the structure can change with a variation in temperature of only a few degrees.

At low temperatures, the forces producing alignment clearly have the upper hand and the molecules thus form an ordered arrangement: this is the *normal crystalline solid state*. The crystal consists of equally-spaced layers of parallel molecules which are in general aligned normal to the plane of the layers and themselves form a regular array (Fig. 9.20(a)).

[1] Very recently, mesophases have been discovered with molecules in the shape of flat discs.

When melting occurs normally in a substance, the order disappears suddenly at the melting point, where thermal motion begins to predominate. When a mesophase exists, the order only *partially* disappears. Several types of intermediate structure are observed.

In the first type, the layers of molecules persist but they can slide over each other, so that the order between any two successive layers is destroyed. Moreover, the molecules within each layer are no longer strictly ordered: their number per unit area remains very close to its value in the crystal, but there is no longer any long-range order. However, some short-range order persists between neighbours, as it does in a liquid (Chapter 7), so that each layer has the structure of what could be called a two-dimensional liquid. This type of mesophase (Fig. 9.20(b)) is described as **smectic**, a word derived from the Greek root σμῆγμα, meaning 'soap', because the consistency of these phases is reminiscent of that of a soft soap.

There is a second type of mesophase, called **nematic** (Fig. 9.20(c)). The degree of order here is less than that of a smectic phase: the molecules are no longer arranged in layers, but they do remain aligned in a single common direction. *The centres of the molecules have completely random positions* as do those of a normal liquid.

Some substances pass from the crystalline state through the smectic phase to a normal isotropic liquid state; others pass through the nematic phase instead; finally, some follow the sequence crystal → smectic → nematic → isotropic liquid. There are even more complex transitions, for instance with several subspecies of the smectic phase occurring one after the other in the same substance or in different substances: we shall not deal with these. We shall, however, be describing later a variety of the nematic phase known as a **cholesteric** because of its important applications.

In the liquids showing mesophases that are known at present, the transition temperatures are very generally in the range 0–150 °C, although a few are known with higher values.

The smectic phase

When a single crystal with the sort of layer structure described above passes into a smectic phase, it is easy enough to imagine the molecular layers in the liquid crystal remaining parallel to the direction they had in the solid. However, in the transformation *from the liquid* to the smectic phase, there is no privileged direction since the liquid is isotropic. In this case, seeds will form at various points and layers parallel to each other will accumulate around each one, but the seeds at different points, and hence their smectic planes, will have different orientations. The structure of the whole mass is clearly going to be very complex, but the geometry of the variation in orientation between different domains can be analysed and used to explain the complicated pictures seen under a polarizing microscope (Fig. 9.21).

An arrangement that is simpler and more interesting from a

Fig. 9.21 Thin section of a smectic phase observed under a polarizing microscope. The variations in tone correspond to regions of different orientation. (M. Kleman, Laboratory of the Physics of Solids, Orsay.)

|———————— 0.5 mm ————————|

fundamental point of view is that of a thin layer (a few tens of μm thick) placed between two parallel glass microscope slides or any other similar glass plates. The smectic phase then becomes optically anisotropic because of the orientation and shape of the molecules: in other words, it becomes doubly refracting (or *birefringent*) like a crystal platelet (hence the name *liquid crystal*). The polarizing microscope is essential for studying materials of this type, since a detailed analysis of the pictures obtained with it enables the structure on a molecular scale to be determined. In fact, it was through the use of this instrument that liquid crystal structures were analysed before the advent of X-ray diffraction techniques. However, the resolving power of the polarizing microscope is limited: it is no more than about one hundred times the size of the molecules themselves.

The importance of the thin smectic layers prepared in that way arises from the action of the plane glass surfaces on the liquid: they produce an orientation of smectic planes making them all parallel to the surfaces throughout the whole thickness of the layer, and this makes investigation much easier[1].

The existence of parallel layers of molecules in specimens with such unique orientation has been revealed by X-ray diffraction. The same technique has enabled the thickness of the layers to be measured and has established its value as *equal to the length of the molecules themselves*. It also shows that the internal structure of a layer is that of a 'two-dimensional liquid' in general, although there are smectics in which short-range order is so significant that the structure of the layer tends more towards that of a 'two-dimensional crystal'. In certain smectics, there is even some correlation between the positions of molecules in successive layers, so that they approach the sort of order to be found in a three-dimensional crystal, albeit a very imperfect one. In these last cases, the phase appears to be almost a solid.

The above examples show how diverse the structures of smectics

[1] Without going into technical details, we should explain that the orientation of a mesophase is always a delicate operation whose success is more often than not due to quite an empirical approach.

can be. We may add that many of them have only been known for a few years, but studies of this sort bring out once again the need for the relatively new notion of intermediate states of order between those of crystals and liquids.

The nematic phase

The type of order still persisting in this phase is the common orientation of the molecular axes. However, the centres of the molecules are completely random in their location, just as they are in a normal liquid, and that is why nematics are much more fluid than smectics. When a nematic phase is produced by transformation of either a smectic or a crystal (Fig. 9.20), the molecular layers are broken up but the molecules themselves remain parallel to each other, just as they were before the phase change. Conversely, when starting from an isotropic liquid, the directional order that exists in the nematic has to be created below the transition temperature: *locally*, groups of molecules become organized into parallel bundles, but different groups have random orientations, *unless a common direction is imposed by some external constraint*.

This is what happens in a thin nematic layer between two glass surfaces, the constraint being provided by the interaction between the molecules and the surfaces as in the case of smectics. We should add that things are far from being straightforward. For instance, if the glass slide is rubbed with filter paper in a single direction, the nematic molecules are then found to orient themselves in that direction, as if guided by the tiny scratches. Other treatments of the glass surfaces lead to molecules taking up directions normal to the surfaces. We can also use the fact that the elongated molecules align themselves along a strong magnetic field, while electric fields have a directive action as well. All these examples show that *the orientation of molecules in a nematic can be controlled by very subtle influences*, a feature which is the basis for applications of this type of liquid crystal.

Physicists amuse themselves by submitting a nematic simultaneously to two contradictory directive influences in order to see what the molecules will do. The answer is known—at least in principle—they choose the most advantageous arrangement from the point of view of energy: in other words, the arrangement that minimizes internal stresses and tensions. For example, if the nematic is placed between two glass slides, one having scratches in a direction perpendicular to those on the other, the structure is twisted, with the molecules in successive planes turning continuously through 90°. Another example would be that of a specimen with molecules *normal* to the glass surfaces being subjected to a magnetic field *parallel* to the surfaces, and so on. These experiments are not just games for the fun of it: not only do the results furnish information about the nature of the molecular interactions, but some of the unnaturally complicated structures are used in liquid crystal devices, for example the figures displayed on watch faces.

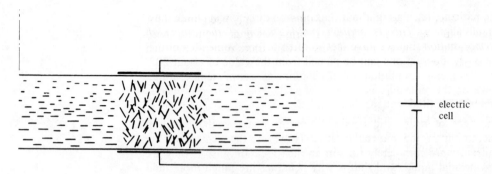

electric cell

Fig. 9.22 Method of forming visible figures on a watch face using liquid crystals.

Such a display is produced by using a small nematic specimen of the type described above in which the molecules are twisted through 90°. In fact, the nematic is placed between two polaroid sheets in an uncrossed position[1]. *The plane of polarization of light passing through the nematic follows the orientation of the molecules,* and so is turned through 90°. Thus the second (uncrossed) polaroid sheet allows no light through, and nor is it reflected. If now small transparent electrodes are placed parallel to but outside the polaroids, and if an electric field is established between them, the molecules are turned as shown in Fig. 9.22. The light can now pass through the uncrossed polaroids and the result is that the part of the arrangement covered by the electrodes becomes transparent (or reflecting). The shape of the electrodes stands out clearly against a black background and if they delineate numerals these will appear as soon as a command is received from the electronic circuits in the watch. The voltage required is only about 1.5 V and *only a tiny electric current flows.* The device can thus function for a very long time with a small electric cell lodged in the case of the watch.

This example shows what can be achieved through the action of an electric signal on the direction of the molecules in a nematic. Using these techniques, as well as many other different ones, any points on the small sheets containing the nematic can be made visible at will. If the signals came from the circuit in a television receiver, then it is easy to imagine the normal cathode-ray-tube screen replaced by a sheet of nematic. Indeed, a lot of the recent research into liquid crystals has been motivated by the hope of producing television screens the size of a post-card and no more than a few millimetres thick. The idea is certainly reasonable enough, but there are numerous technological problems that have not yet been overcome.

Fluctuations in the orientation of nematic molecules
In the same way that the atoms in a crystal are continually vibrating about their mean positions, so the axes of the molecules in

[1] A sheet of polaroid allows a maximum intensity of light to pass through it when the light has a certain plane of polarization. A second sheet, *not* crossed with the first, would allow the light to pass as well, but when the second sheet is turned through 90° in its own plane (that is, it is in a 'crossed' position relative to the first), the light is completely stopped.

a nematic are oscillating about the common direction in which they are all aligned. *This is another manifestation of the universal thermal motion* that we have met so often before. Some idea of the size of the fluctuations can be gained from the value of about 20° for the mean amplitude of oscillation at room temperature. However, the molecules are so tightly packed that their movements cannot be entirely independent and, for that reason, a collective oscillatory motion is preferred: in other words, groups of molecules oscillate together over regions that are quite large compared with the size of a single molecule (the regions are about 1 μm across). This angular oscillation or rotation of molecules affects the propagation of light because it produces variations in the refractive index. The regions are large enough for the molecular motion to be directly visible and it appears as a continuous swarming of bright points that is characteristic of all nematics (Fig. 9.23). It was a phenomenon that proved highly intriguing to those who first observed it.

It is worth pointing out that this effect is really quite remarkable: it is only through a conjunction of favourable circumstances that, in this special case, we can observe something that is normally hidden from us and yet is one of the essential features of atomic structures: *the unceasing thermal motion of atoms*, which we cannot, and never shall be able to, influence in any way. It is also a reminder that all the structures we propose with fixed atomic positions are only averages: each of them should be accompanied by a warning notice: 'Take care: the atoms are not really fixed in the positions indicated but are always vibrating about them.'. Indeed, in some crystal structures of high accuracy, the mean position of each atom is surrounded by an ellipsoid to indicate the amplitudes of vibration in different directions relative to the crystal axes.

Lines of defects in nematics: disclinations
When a thin layer of nematic has no overall fixed orientation, the change in direction of the molecular axes is neither continuous

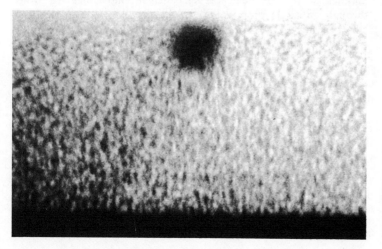

Fig. 9.23   A sheet of nematic seen under a polarizing microscope. The granular appearance is due to fluctuations in the orientation of molecules. The image is constantly moving because of thermal agitation. (M. Kleman, Laboratory of the Physics of Solids, Orsay.)

nor progressive, but instead gets concentrated in narrow regions around 'defect lines'. This is very similar to the way in which the irregularities in atomic positions in a crystal concentrate around a dislocation line (Chapter 5, p. 111). By analogy, the lines of concentrated defects in nematics are called 'disclinations' because it is the inclination of the molecules rather than their location that is involved here.

The directional disorder in the immediate neighbourhood of a disclination line is large, and it results in a strong scattering of light. The lines then appear under a microscope as if they were black threads. This is very characteristic of nematics and in fact gave them their name (Greek νῆμα, 'thread'). These threads do not correspond to any foreign substance occurring in the material but rather to a sort of fold or twist in the structure that is capable of being displaced or deformed. Our knowledge of liquid crystals has thus reached the stage where we have established not only the general structural models but also the nature of local defects.

The cholesteric phase

As in smectics and nematics, the molecules giving this type of mesophase have a rod-like shape—elongated and rigid—but they also have another quite special property: *chirality*. This means that they are asymmetrical in such a way that they cannot be made to coincide with their image in a plane mirror by simple movements such as displacement or rotation. It is like a right hand, say, which can be considered as the image of a left hand but will not fit into a left-hand glove. The molecule illustrated in Fig. 9.24 is an example of a chiral molecule.

In a sheet of perfectly oriented nematic, the molecules have their axes parallel to the walls of the glass slides and thus have the same direction everywhere: those in any layer are parallel to those in the neighbouring layer. In a cholesteric, however, the asymmetry of the molecules causes those in one layer to make a small angle with those in the adjacent one. The process is repeated from one layer to another, so that in the whole specimen the molecules are turned through an angle proportional to the distance h from the solid surface (Fig. 9.25). The structure is therefore *helical*, and is characterized by the distance between two layers that have the same direction (that is, between which the molecules have rotated

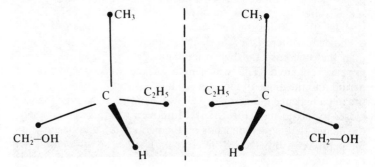

Fig. 9.24 Example of a chiral molecule. Two enantiomorphic molecules (one 'right' and one 'left'), which show reflection symmetry with respect to a plane and *cannot be superposed.*

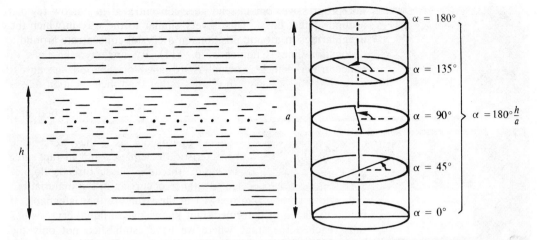

through 180°). The orientation of molecules is determined here only by the direction of their axes and not by the sense. Even if the head and tail are different, it is quite immaterial which is pointing in which direction, so that the molecules recover the same orientation after only a half-rotation through 180°. The process is repeated over the whole thickness of the specimen, so that the cholesteric is characterized by a periodicity in a direction normal to the plane containing the axes of the molecules. The period, denoted by a in Fig. 9.25, varies from one substance to another with a typical order of magnitude of 1–2 μm and, for a given substance, it varies with temperature.

In short, this mesophase is thus a particular variety of nematic. Its name originates from the fact that the first examples were encountered in derivatives of cholesterol, which everybody knows as the substance that must not be present in blood in excessive amounts. Nowadays, cholesterics with quite different molecular formulae are known.

The helical structure gives cholesterics very special optical properties. First of all, their optical rotatory power is enormous: several thousand degrees per millimetre or one hundred times greater than that of common optically-active solids and liquids. In addition, the periodic layer structure of the sheet arising from the helical twisting of the molecular axes enables it to act on a beam of light just as the periodic layers of atoms in a crystal act on an X-ray beam. Although the two phenomena are comparable, their scales differ by a factor of around one thousand, both for the wavelengths and for the periodicity of the layers. For a given angle of incidence, there is a wavelength which is reflected much more than the others, so that when illuminated by white light, the sheet appears coloured. The wavelength picked out depends on the period and the angle of incidence. Thus, a sheet appears blue under grazing incidence but green or yellow when the angle of incidence is closer to the normal (Fig. 9.26). Moreover, with a given geometry, the colour that is seen depends on the periodicity, and hence on the temperature, with which it therefore varies. *The colour change is in fact quite appreciable with a variation of only 0.1 °C.*

Fig. 9.25 Illustrating the orientation of molecules in a cholesteric phase. Molecules are represented by short lines of a length proportional to their projection on to the plane of the page. The directions of their axes are indicated in the diagram on the right.

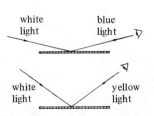

Fig. 9.26 Variation of the colour of reflected light from a sheet of cholesteric material as the angle of incidence changes.

Fig. 9.27 A thermometer consisting of a sheet of cholesteric material. The photographs were taken at 15 °C and 21 °C.

This property is used to measure the temperature of a surface by laying a film of cholesteric on it. If the temperature varies from one point to another, the colours that are visible yield a sort of temperature map. This is the principle behind *thermography*, which is used in medicine for the detection of sub-cutaneous tumours.

A *liquid crystal thermometer* consists of a band along which a cholesteric material is spread out from one end to the other. The material is such that its periodicity varies uniformly with distance along the band. The section which reflects light changes with temperature, which can then be picked out quite easily. Figure 9.27 shows an example of such a thermometer covering the range from 10 °C to 30 °C.

Liquid crystals provide remarkable examples of advances made in our knowledge of the structure of matter only very recently, but achieved, nevertheless, using relatively simple and well-known experimental techniques. In addition, no sooner had the mechanisms of molecular orientation been successfully analysed in detail and the means of changing it at will been found, than technological applications were being developed.

Soaps and lipids

The systems we are going to examine in this section are mixtures of water with substances whose molecules have one special feature: they are elongated and possess a head and a tail having different properties. *The head is polar*: in other words, it behaves like an electric dipole or a polar group which, we remember from p. 77, consists of equal positive and negative charges with separated centroids. The head is said to be *hydrophilic* because it readily links up with water molecules or other polar groups. *The tail consists of one or two paraffin chains and is non-polar*. Unlike the head, therefore, it does not favour contact with water molecules but prefers other non-polar groups and is thus called *hydrophobic*. The tail is also *flexible* so that the two ends of the molecule can behave quite independently. Two such molecules can be linked ·head-to-head or tail-to-tail, but never head-to-tail.

Alkaline soaps form one important group of such substances. They are sodium or potassium salts of saturated fatty acids, with formulae that enable the head and tail to be clearly distinguished:

$$CH_3-(CH_2)_{n-2}-COOM$$

hydrophobic   hydrophilic
tail   head

where M stands for K or Na, and n is an even number between 8 and 22.

Lipids are another enormous and diverse class of such molecules, often consisting of fatty acids: they are found among the constituents of living matter. As an example, we give the formula of *lecithin* (phosphoglyceride), which occurs in the cell membranes of animals:

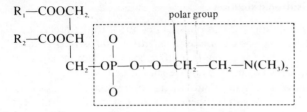

The hydrophobic tail consists of two paraffin chains R_1 and R_2.

Phases in the water–soap system

The reason for dealing with soaps in this chapter is that they provide yet another example, like those of the preceding sections, of structures intermediate between order and disorder. Such structures are of interest and importance because they determine the properties of many materials in everyday use and because they probably occur in many biological materials, although in this area our knowledge is still quite limited. The subject is certainly extremely complicated, so that the account of soaps which follows is neither complete nor detailed.

The state of a water–soap mixture is determined by two parameters: the *temperature T* and the *concentration c* (the ratio of the mass of soap to the total mass of the mixture). The structure of the system does not change continuously as a function of T and c: over a certain range of both, the structure remains of the same type so that it is said to be in the same *phase*. Outside these ranges, variations of temperature at a fixed concentration or changes in the water content at constant temperature can make the system pass abruptly from one phase to another.

Experiments show that there are large numbers of different phases both for these systems and for lipids. Their structures have not all been completely unravelled, since it is only in the last thirty years or so that fundamental work with adequate experimental techniques has begun, even though the physical properties of soaps have been known for a very long time because of their technical applications.

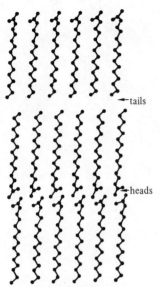

Fig. 9.28 Crystal structure of a soap derived from pentadecyclic (pentadecanoic) acid seen in projection on to the plane containing the molecular axes. Notice that the molecules are in head-to-head or tail-to-tail contact.

We shall begin with the two extreme phases which are also the simplest: one corresponds to the ordered state, the normal crystal, and the other to the disordered liquid state.

The crystalline phase

The conditions that apply in this case are that the soap is *anhydrous* and is *at or below* the ambient temperature. The molecules, being elongated, pack together parallel to each other, as we should expect, and form a 'good' crystal without any irregularities. All the molecules are stretched to their maximum extent and, as we have already indicated, have such a directional sense that head-to-tail contact between neighbours is avoided. The crystal structure of a typical soap is illustrated in Fig. 9.28 and shows how all these features are incorporated.

The isotropic solution

The conditions for this phase, in contrast to the preceding one, are that the soap *has a low concentration and is at a relatively high temperature*: for example, a 30% soap solution in hot or very hot water. This is simply what we should call 'soapy water' and, as we all know, it is quite fluid but has one special feature: it is always slightly cloudy and appears bluish if viewed in sufficient depth. These aspects arise from the scattering of light (p. 54) and mean that the solution must contain particles that are larger than the molecules themselves. Solutions in which the molecules are perfectly free, such as sugar, are quite clear because the particles are not large enough to scatter light of visible wavelengths.

Thus, the idea arises that *molecules of soap are agglomerated into the so-called* **micelles**. Since we know that the hydrocarbon tails are hydrophobic and tend to avoid contact with water, we are led to the following model for a micelle. A globule is formed with a surface completely covered by the polar heads and the paraffin chains folded inside this protective skin in some disorder (Fig. 9.29). Given the length of the molecule and assuming the globule to be spherical, its diameter would be of the order of 4–5 nm. However, there is little direct evidence to confirm the shape of the micelles and elongated cylinders with the same diameter or even flat micelles are thought to occur. *The micelles are dispersed at random in the homogeneous solution* and, judging by the fact that alkali ions exist in the solution, it is assumed that the micelles are left negatively charged. This causes them to repel each other, thus preventing coagulation and stabilizing the solution.

4 nm

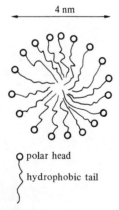

○ polar head

〈 hydrophobic tail

Fig. 9.29 The structure of a micelle in soap.

The intermediate phases

Quite a large number of phases have been found between the crystalline state and the disordered solution in which the order is only partial. The properties of these phases and the conditions under which they are stable depend on which system is being considered and we describe only a few typical structures without giving precise details of their production.

Intermediate phases in anhydrous soap

First of all, we consider how the structure of anhydrous soap changes with temperature as it passes from the crystalline solid at lower temperatures to the molten isotropic liquid at higher ones. Between these two states, there are several mesophases of the *smectic* type, optically anisotropic and having a pasty consistency. In sodium stearate, for instance, five different phases can be detected between room temperature and the melting point of 350 °C. We limit ourselves to a description of one of them with a structure as illustrated in Fig. 9.30 which is in agreement with X-ray diffraction data. It consists of straight bands, perpendicular to the page, formed from two layers of molecules with head-to-head contact. Between the bands, the paraffin chains forming the tails are in disorder, rather as they are in a melted paraffin. In this phase, the order arises from the regularity of the arrangement of bands and is only two-dimensional. Moreover, the order only extends over small regions, each of which has a random orientation in relation to the others, rather like the crystallites in a polycrystalline solid. The structure is peculiar in that the quasi-liquid disorder in the paraffin chains is combined with the long-range order in the bands of molecular heads.

Even if we have begun to get some idea of the structures of these mesophases, we are not in a position to explain the origin of the forces that produce the stability of any particular example, each so complex in appearance. Nor do we know why several phases with slightly different structures should succeed each other over small temperature intervals.

Gel and curd

We now consider a series of mixtures of crystalline soap with an increasing proportion of water, but remaining at a constant temperature, that of the surroundings. (In fact, a mixture of given proportions is prepared by homogenizing it at a higher temperature and then allowing it to cool.) With a high enough proportion of water, an isotropic solution of the micelle type already described is obtained. For smaller proportions, the result is a 'gel': a soft translucent mass with quite an elastic consistency, such as *soft potash soap*.

The gel is also a *mesophase with a structure of the smectic type*. The molecules form parallel and equidistant sheets by alternating and overlapping in such a way that the polar heads are all on the outer surfaces (Fig. 9.31(a)). Within each sheet, the molecules form a regular two-dimensional lattice (Fig. 9.31(b)), while layers of water separate the sheets so that adjacent ones can slide over each other. The relative positions of the sheets are well defined in a direction normal to their planes but not at all parallel to them. Given the weakness of the bonds between the sheets, it is not difficult to see that such a structure would produce a soft, easily-deformed material. What *is* astonishing, and still unexplained, is the uniformity of the spacing.

In some hydrated soaps, the gel phase is unstable and is replaced by a *curd*, which is a mixture of crystalline soap and soap solution.

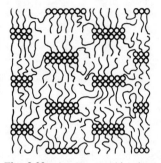

Fig. 9.30 A mesomorphic phase in anhydrous soap. Bands of molecules perpendicular to the page consist of ordered arrays with head-to-head contact, while the tails are in disorder.

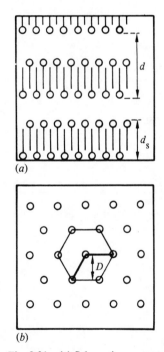

Fig. 9.31 (a) Schematic representation of the structure of a gel of potash soap. The molecules overlap in opposite senses to form sheets of thickness d_s which varies from 1.85 nm for soaps with C_{14} to 2.85 nm for those with C_{22}. The repeat distance between two sheets, d, is about 3 nm. The space between the sheets is occupied by water molecules. (b) The regular array of molecules in one sheet.

The resultant material is quite hard and opaque and forms the basis of household and toilet soaps.

Neat soap and middle soap

In the course of the industrial manufacture of soap, the soap–water mixture is taken to temperatures of 150–200 °C. Soap-makers have noticed for a long time that at these temperatures the mixture has different physical properties according to the proportion of water. At about 30% water, *neat soap* is formed, quite fluid and easy to work, but with greater amounts (50%), *middle soap* is formed, which is very viscous and difficult to handle.

These are two quite distinct mesophases. Neat soap is a *smectic phase with parallel sheets* having water molecules between them. Unlike the gel described above, however, the soap molecules are disordered within the sheets (Fig. 9.32(a)) even though the distance between sheets is quite uniform. In middle soap, the molecules form cylinders with the heads localized on the outer surfaces and with water between them. *These cylinders are arranged in a hexagonal lattice* (Fig. 9.32(b)). There is thus long-range order in both types, but it is two-dimensional in middle soap and only one-dimensional in neat soap. The fluidity and softness of neat soap arises from the ability of the molecular sheets to slide over each other, while the other phase, in spite of its greater concentration of water, is more rigid because of the order existing between bundles of cylinders formed from soap molecules.

These results, taken as a whole, are clearly very complicated, yet we have still only given a fragmentary description of the soap–water system. Not only that, but the soap molecule is simple compared with those of many lipids, so that the complexity of most of the other systems can well be imagined.

States of intermediate order form a difficult field of research because current methods of elucidating the structures cannot yield complete solutions owing to the complex mixture of order and disorder. However, although it is as yet relatively unexplored, it is a field that could become more and more important in our understanding of the atomic structures of a large number of substances. There are profound reasons (see later on p. 226) to

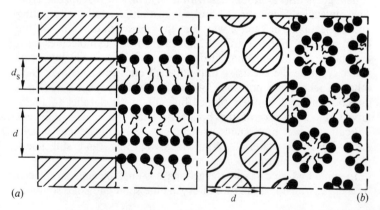

Fig. 9.32 (a) Layer structure of *neat soap* with regular spacing of sheets. (b) The structure of *middle soap*. The cylinders of soap molecules form a uniform hexagonal array.

(a)

(b)

think that states of intermediate order must be quite common, even perhaps normal, when the constituent molecules are large and have complicated shapes, and it is almost always from such molecules that the organs of living matter are constructed—or at least parts of the organs, such as cell membranes. That is the reason why the ideas that have been developed in the patient study of relatively simple systems—of, for example, liquid crystals and soaps— will perhaps find their most important and novel applications in biology.

10 Composite materials, suspensions and colloidal solutions

Until now, we have been mainly concerned with constructing, from atoms or molecules, models of structures that we know to exist under various conditions of temperature and pressure. The descriptions of these models have been, in effect, a summary of all the experimental work carried out on the determination of atomic structures over the last sixty years. What we have called a 'phase' is identified or defined only by its structure and not at all by its extent: the same phase can occur both as a tiny grain of sand or as a large quartz crystal, both as a drop of water or as the contents of a lake.

However, the structure of matter can be described by taking a different approach. We start with things as they are, the objects that are all around us, and analyse them, say under an optical microscope, into a collection of contiguous regions, each of which appears to be homogeneous. However complex any system or any object may be, it is made up from a collection of phases as described in earlier chapters: crystals, amorphous substances, liquids, etc. The properties of the system as a whole do not depend only on the atomic structures of the constituent phases, but also on the sizes of their domains and how they are arranged in relation to each other.

This method of analysis can be carried out with inanimate matter, such as rocks, metals or various synthetic products, but if attempted with natural substances or living matter, the results are not nearly so clear-cut. For one thing, their organization is so complex that we have to come down to a much smaller scale to find 'homogeneous' regions. Moreover, such matter is found to contain many phases with intermediate states of order such as we studied in the last chapter, phases which have ill-defined boundaries and, it must be admitted, which also have structures that are still not very well understood.

Composite materials

In this section, we limit ourselves to *composite materials*, by which we mean solid systems consisting of well-defined and distinct phases. The essential feature of such materials is this: even if all the properties (mechanical, electrical, etc.) of the various phases are known exactly, those of the composite material cannot be found simply by taking an average over the constituents and allowing for their relative proportions. The material as a whole has its own individuality arising from *interactions between adjacent phases* and from the special properties of the boundaries or interfaces between the phases.

Nearly all common solids, natural and synthetic, come into the category of composite materials, so that they are clearly worth studying as a group. However, they are also of importance because their properties may be more interesting than those of the pure constituent phases. We have already met several examples of this in connection with crystalline solids (p. 134) where all the phases were crystals. We now extend these cases to mixtures of crystals with amorphous substances.

When two phases with different structures are in contact along a grain boundary, there is cohesion between them because there are attractive interatomic forces between atoms in *different* grains as much as there are between atoms in the *same* grain. A simple experiment that demonstrates these forces occurs in an instrument commonly found in mechanical workshops, known as a slip gauge or block gauge, and containing wedges with perfectly plane and clean surfaces. If two of these surfaces are placed in contact, they are attracted so strongly to each other that they cannot be separated by pulling but can only be made to slide (Fig. 10.1). These atomic forces decrease rapidly with distance, however, and a thin layer of grease or a few particles of dust on the surfaces would be enough to destroy the cohesion between the wedges.

It is difficult enough in a crystalline phase to calculate the binding energy knowing the atomic positions and electronic configurations. So it can easily be seen that it would be almost impossible to calculate the cohesive forces between two phases with different structures, where there are atoms along the interface in slightly perturbed positions (see Fig. 6.7). The theory of interfaces is not yet very well developed.

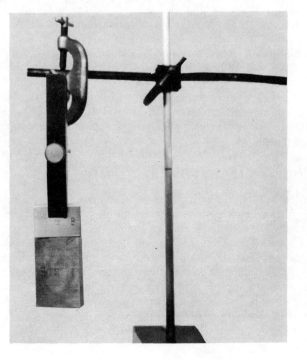

Fig. 10.1 Attraction between wedges of a slip gauge. The large wedge of mass 180 g is supported solely by the cohesion between perfectly plane and clean surfaces.

The cohesion between grains can be so strong that, when the solid is fractured, the break occurs in the interior of the grains rather than between them. However, there can be loss of cohesion *between* grains in some cases, particularly if impurity atoms have collected along the boundary and weakened the interatomic forces between grains (Fig. 10.2).

When the cohesion between two phases with different structures is very strong, the atoms of the harder one produce a small amount of deformation in the other, so that interfaces are created where the structures are distorted. We have already seen that such interaction between phases explains the hardening of tempered steel, in which the second phase, martensite, occurs in the form of needles (p. 121).

Other metals find their strength increased in the same way by inclusions, not so much because of the intrinsic properties of these other materials, but because they form strong bonds with the metal and because the interfaces stop the movement of dislocations which causes the plastic deformation of the metal (p. 117). The strength of a metal such as steel or aluminium is increased by inclusions consisting of fine particles of oxides or nitrides. The effect depends not only on the number of inclusions but on their shape as well. We can illustrate this by the example of *grey cast iron*, that is, iron containing a second phase, graphite. Although it may contain the same proportion of carbon, the iron is brittle when the graphite is in the form of irregular flakes and ductile when it consists of spherical granules (Fig. 10.3).

Two phases that are strongly bound together in a composite material can be deformed in different ways when large external forces are applied as long as the whole material does not disintegrate. This sometimes enables us to benefit from the addition

Fig. 10.2 View under a scanning electron microscope of the surface revealed by an intercrystalline fracture. The weak cohesion between grains has allowed them to separate from each other. The sample is an industrial titanium alloy. (P. Lacombe, Metallurgical Laboratory, University of Paris-Sud.)

1 μm

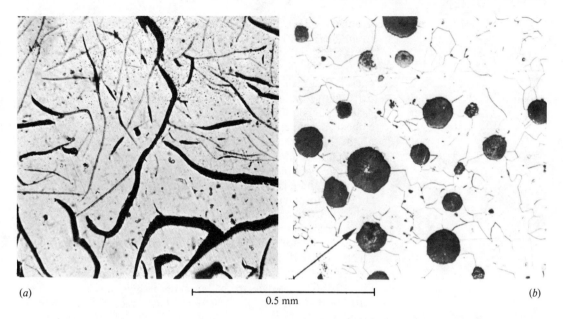

(a)

0.5 mm

(b)

of quite different and contrary properties of the two portions. We give a few examples of this.

Extremely hard tools are needed to turn metals on a lathe quickly and precisely. Tungsten carbide is one of the hardest materials available but the carbide itself could not be used on its own to fashion a tool of the desired shape. A composite material is used instead, in which grains of the carbide are 'cemented' together by cobalt, a quite ductile metal. In a similar way, high speed rotating discs, capable of cutting the hardest materials, are made by including diamond powder in copper.

Fibrous composites

Some solids can be produced in the form of fine fibres with very large tensile strength but with the big disadvantage of great brittleness, which limits their usefulness. They show little resistance to sharp impacts because there are always small scratches and cracks on the surface which are sources of weakness and which enable fracture to occur readily, just as a small cut at the edge of a sheet of strong paper makes it easy to tear completely. If a bundle of such brittle fibres is embedded in a thermosetting resin (p. 191) which adheres well to it, this more flexible material protects the fibres from fracture through impact, but allows them to preserve their tensile strength. The composite material is thus much stronger than the plastic alone, at least in one direction, and much less brittle than the fibres alone.

The most commonly-used fibres are either of spun glass, which is easy to produce, or of asbestos, although the best of all are carbon fibres, whose extraordinary strength we have already indicated (p. 185). For a given weight, composites of fibre and plastic are among the strongest materials known (they are used in satellite

Fig. 10.3 Optical micrographs of two specimens of cast iron consisting of two phases, ferrite (an iron–carbon solid solution) and graphite. (a) The graphite is in the form of fine irregularly shaped flakes: the cast iron is brittle. (b) By adding a little magnesium, the graphite takes the form of spherical granules with diameters of a few hundredths of a millimetre: the cast iron is then ductile. (P. Lacombe, Metallurgical Laboratory, University of Paris-Sud.)

re-entry cones). However, even if the principle behind such strengthening is simple, we should appreciate that its development has needed, and still needs, a great deal of work.

The disadvantage of common plastics, even the thermosetting ones, is that their use is limited to temperatures up to about 250 °C only, above which they begin to decompose. To achieve higher usable temperatures with composite materials, the fibres are embedded in metals which, like the plastics, protect the surfaces of the high strength fibres and eliminate their brittleness. Pure aluminium, reinforced by amorphous silica fibres, *has a greater tensile strength than the best light alloy* (Fig. 10.4).

Finally, there is a rather curious process for creating very strong fibres by forming them naturally in the body of a metal. Some melted alloys with certain well-defined compositions solidify in two regularly alternating phases. These are called *eutectics*. Conditions for the cooling of one such eutectic system have been found under which one of the phases crystallizes in the form of long parallel needles of about 1 μm diameter, which are regularly spaced as in Fig. 10.5. The needles are of tantalum carbide and are extremely hard, so that the composite material has a remarkably enhanced strength.

In quite another area, there is one natural material which is comparable to a fibre composite: *wood*. The fibres are made of cellulose chains (p. 181): they are embedded in a rather soft amorphous material called *lignin*. Everybody knows that a log is strong in the direction of the fibres, but splits easily lengthwise because the lignin is easily torn and the fibres become separated.

Fillers in plastic materials

Many plastics are not used in a pure state, but have 'fillers' added to improve this or that property a little. Such additions must not be

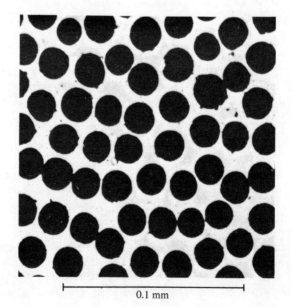

Fig. 10.4 Cross-section through a composite material: fibres of silica in aluminium. (J. W. Martin, *Strong Materials*, Wykeham.)

0.1 mm

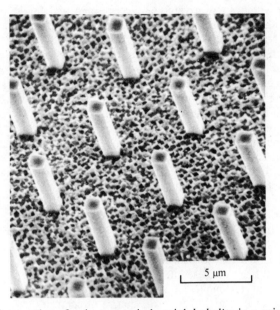

5 μm

Fig. 10.5 Scanning electron micrograph of a composite formed spontaneously by a special technique of solidification. The *fibres* are crystals of tantalum carbide in a *matrix* of nickel, cobalt, aluminium and several other materials. The fibres have been revealed for the micrograph by using a selective chemical action to dissolve the matrix. (H. Bibring, Office National d'Études et de Recherches Aérospatiales, Chatillon-sous-Bagneux, France.)

expensive so that, for instance, industrial *bakelite* is a mixture of the resin and of wood dust or mineral powder. The volume of the material is clearly increased for a given amount of resin and this can make the object more convenient to use, but, *what is more*, the mechanical properties are improved.

Some fillers play an essential and very beneficial role. This is the case with *carbon black in rubber*. If vehicle tyres are to stand up to wear and tear, carbon black is indispensable and its quality is a decisive factor in that of the tyre. There must clearly be a strong interaction between the rubber and the small carbon particles, but the mechanism is not yet clear.

Concrete and mortar

These materials, so essential to the building and construction industries, can clearly be seen to have a heterogeneous structure on a scale of millimetres or centimetres, and it is precisely this mixing of phases that gives them their special properties.

Fresh concrete is made from a mixture of cement, water and aggregate: cement is a synthetic product consisting mainly of calcium silicates and aluminates, while the aggregate is a mixture of sand and stones up to a few centimetres across, although only fine sand is used in mortar. Fresh concrete can be poured, and is fluid enough to fill completely a mould of any shape. After a few hours, it begins to harden and within days a sort of 'artificial rock' is obtained having the shape of the mould, which can then, of course, be removed. The rock-like concrete consists of the larger stones of the aggregate bound tightly by the hardened paste of cement in an intimate mixture with the smaller grains of the aggregate.

On a finer scale, this hardened paste is found to be a tangled mass of small crystals: mixtures of hydrated calcium silicates, hydrated calcium aluminates, gypsum and several other compounds.

On the whole, these crystals are very small and for a long time it was thought that hydrated cement contained an amorphous phase. No such phase exists, however, and the scanning electron microscope gives very clear pictures of faceted crystals (Fig. 10.6). The essential feature of concrete is *the strong adhesion of these crystals to each other and to the grains of the aggregate*. When completely hard, a block of concrete is capable of withstanding a considerable pressure without breaking (of the order of 40 MN m^{-2}). However, when in tension, the stresses it can withstand are ten times smaller. Under compression, the small crystals remain stuck to each other, but if the external forces are such as to stretch the concrete, they can very easily be detached from each other or even broken.

Thus, the important properties of concrete depend on the cohesion between crystals: on the atomic scale, everything relies on the interaction between surface layers in contact and in particular on the surfaces of hydrated crystal structures into which the free water has been absorbed. Unfortunately, as we have said, the theoretical basis of these interactions is not very well developed, so that current techniques are the result of a long succession of empirical trials undertaken in the course of centuries to satisfy the needs of engineers and builders. Today we can make sense of the techniques using our knowledge of atomic structures, and modern analytical methods are indispensable in maintaining consistency in the properties and in continually producing improvements in them.

To overcome the relative weakness of concrete under tension, a still more complex composite material is used: *reinforced concrete.* A steel rod of 1 cm diameter can withstand with complete safety the same *tensile* force as a concrete post with a diameter of 15 cm. However, the concrete post can support a weight of 70 tonnes without being damaged, a load that would cause the thin steel rod to buckle. The idea therefore arose of introducing the metal rod into the concrete and benefiting from the two advantages at the same time. This is a practical possibility because the coefficients of volume expansion of steel and concrete are very similar so that *there is excellent adherence between them at all temperatures.* Nowadays, the part played by reinforced concrete in the construction industry is well known to everybody: it has the advantage of being far less expensive than the metal alone for the same strength and in addition is not subject to atmospheric corrosion as is the steel. The main disadvantage is its weight compared with that of the metal alone, and for that reason its use is confined to fixed structures.

Finally, there is the additional advantage of using *prestressed concrete.* Here, the steel rods are held under tension while the concrete is poured round them. When the concrete has completely hardened, the tensions are released so that the whole reinforced structure is under a strong compressive force, which is so large that the concrete remains under compression rather than tension even in extreme conditions of operation. The risk of cracking is greatly reduced, and larger structures such as vaults or long bridges can be made considerably lighter.

1 µm

Fig. 10.6 Electron micrograph of hardening concrete one day after pouring. The tangle of needles are hydrated calcium silicate crystals. (M. Regourd, Centre d'Études et de Recherches sur les Liants Hydrauliques, Paris.)

Dispersed phases

Under this title, we shall be looking at a variety of substances with one feature in common: they are all mixtures of two phases in which one is a continuous fluid and the other is in a finely dispersed state, by which we mean that it consists of particles whose dimensions are of the order of 1 μm or less. We have seen that interfaces between phases already play an important role when the particles of one phase are considerably larger than 1 μm. As their size decreases, however, the ratio of their surface area to the total volume increases (e.g. for a sphere it is $4\pi r^2/(4\pi r^3/3)$ or $3/r$), so that for very small particles indeed it is so large that the interfacial actions could predominate. At some point, even the apparently obvious distinction between a system consisting of one homogeneous phase and one consisting of a mixture of two phases becomes blurred. There is a continuous transition from what is regarded as a 'host' fluid phase having particles immersed in it to a true solution where the particles have become isolated molecules. The intermediate state where the distinction becomes difficult is called a 'colloidal state'. Examples occur in a variety of forms depending on the state of the material in the host phase and the dispersed phase: we have suspensions of solids in a liquid or gas, emulsions of two liquids or of a gas in a liquid, and so on.

Examples of systems that correspond to such a description are not uncommon: indeed, we encounter them every day, whether in our natural environment (mists and fogs, clouds, dust in the air, muddy water in a stream, etc.) or among the ingredients used in cooking (milk, mayonnaise, white of egg, whether raw, cooked or beaten stiff, etc.). Most of these systems appear homogeneous when we look at them, yet they cannot be classified along with any of the simple states of matter that we have examined previously.

The specific structure of each phase present in the mixture is already known to us and has nothing exceptional about it. The problems posed are all concerned with the *interaction between phases* and with the *role of interfaces*. Leaving the subject of interfaces until a little later, we turn first to an aspect of the interaction between phases: why it is that mixtures of the sort we have described should appear to be stable. The explanation of this involves several different phenomena which we introduce in turn.

Study of a collection of particles immersed in a fluid medium

Sedimentation of a dispersed phase
Consider a spherical particle of radius r and density ρ immersed in a fluid medium of density ρ_0. The forces acting on the particle are its weight $(4\pi r^3\rho g/3)$ and the Archimedean upthrust $(4\pi r^3\rho_0 g/3)$, giving a resultant downward force of $4\pi r^3(\rho - \rho_0)g/3$. Thus, if the particle is more dense than the fluid, it will fall and be deposited at the bottom of the containing vessel (for example, mud at the bottom of a pond). If the particle is less dense, it will rise to the surface (for example, cream on the top of milk). It would seem to follow from this that the particle is only in equilibrium when the two

densities are equal, yet in fact the system that we observe is not in an equilibrium state, as we shall see.

Speed of sedimentation

Suppose we start from a system made homogeneous by prolonged and vigorous agitation. If it is then left to itself, sedimentation will occur as predicted in the previous paragraph—but with what speed? Now as well as the weight and upthrust, which are always present whether stationary or not, a particle that moves in a fluid experiences an additional force due to the viscosity, a force which is proportional to the speed. As the particle sinks or rises, it very quickly reaches a limiting speed, u, where the resistance to its motion exactly balances the accelerating force. The calculation is quite simple: a spherical particle of radius r in a fluid whose coefficient of viscosity is η experiences a viscous drag equal to $6\pi\eta r u$. The limiting speed (also called the *terminal velocity*) is thus given by the equation

$$6\pi\eta r u = 4\pi r^3 (\rho - \rho_0) g / 3$$

The limiting speed is thus proportional to the difference between densities and naturally becomes zero when they are equal. More important, the speed is proportional to the square of the radius and therefore *decreases sharply as the particle becomes smaller.*

At this stage, it is important to look at some numerical values, so in Table 10.1 we give two examples: a drop of water in air, and a particle of a light solid (relative density 2.0) in water. It illustrates that, on the scale of our everyday experience, *particles of diameter less than 1 μm in water seem to be in equilibrium because they are practically stationary.* It is the same for drops of water in air if their diameter is less than a few tenths of a μm, but here another phenomenon is involved.

Until now, we have considered the fluid phase to be stationary, whereas in practice it is disturbed by convection currents arising from small inequalities in temperature. In particular, even when it is free and apparently quite calm, the atmosphere contains air currents with speeds of at least several centimetres per second. The limiting speeds of Table 10.1 are speeds *relative* to the fluid phase and if they are smaller than those of air currents, the latter simply move the particles around, as can be seen with dust particles dancing in a beam of light. The drops of water in a cloud are

Table 10.1   Sedimentation speed of spherical particles.

Radius (μm)	Limiting speed of fall	
	Spherical particles of relative density 2.0 falling in water	Water droplets falling in air
100	2.2 cm s^{-1}	1.1 m s^{-1}
10	0.2 mm s^{-1}	1.0 cm s^{-1}
1	0.8 cm h^{-1}	40 cm h^{-1}
0.1	2.0 mm day^{-1}	10 cm day^{-1}
0.01	0.02 mm day^{-1}	1 mm day^{-1}

another example: even if they have diameters of several μm they do not fall but are carried along by winds or atmospheric currents.

Returning to our equation, we see that the speed of sedimentation, u, is also proportional to the acceleration due to gravity. If the suspension is placed in a *centrifuge*, the gravitational force is replaced by an apparent centrifugal force which tends to carry the heavy particles as far as possible from the axis of rotation and thus to the bottom of the containing vessel. A rotational speed of 3000 revolutions per minute produces an effective gravitational force at 10 cm from the axis of about 1000g, so that sedimentation occurs in a reasonable time even if the particles are very small and the difference in densities is slight. This is how cream is separated from the whey in making butter from milk.

The real equilibrium of a suspension: the influence of thermal agitation

All particles, like all atoms and molecules, are subject to the universal thermal motion. *The average energy associated with this disordered movement is proportional to the absolute temperature and is the same for all free particles, whatever their nature or their mass.* The motion is directly visible as *Brownian motion* for very small particles (p. 28) and its existence is also demonstrated indirectly by *diffusion* (p. 28) which, as we have seen, implies the movement of molecules in the body of a fluid.

The effect of this thermal motion is that the equilibrium state of a suspension is not exactly as we first described it: the particles do not get separated out on the top or in the bottom of the container, with a sharp boundary between the pure fluid and the dense mass of accumulated particles. Such an abrupt discontinuity is replaced by a diffuse layer, *where the number of particles per unit volume increases gradually from zero to its maximum value.* The profile of concentration can be calculated exactly, but the important parameter is the thickness or height H of the transition zone. As a rough measure of this, we can use the fact that the work needed to raise a particle through the height H (i.e. mgH) is approximately equal to the mean thermal energy kT (p. 38). Table 10.2 gives numerical values of H for different particles at room temperature.

As far as air molecules are concerned, the table shows clearly that the height of a room with a ceiling at three metres is negligible compared with H, so that the density or pressure of air throughout the room is constant. In contrast to that, the pressure varies

Table 10.2 Height H of the transition zone for the sedimentation of particles due to thermal motion.

	Radius	Height H
Oxygen or nitrogen molecules at 20 °C	0.4 nm	30 km
Particles of relative density 2.0 in water at 20 °C	5 nm	180 cm
	10 nm	22 cm
	50 nm	1.8 mm
	100 nm	0.2 mm

significantly when passing from sea level to the summit of Mont Blanc. A mixture containing particles of relative density 2.0 in water is only stable over a significant height if the diameter is less than 50 nm, but since Table 10.1 shows us that in this case the speed of sedimentation is extremely slow, equilibrium is never in practice reached.

Thus, a suspension of fine particles in water can be preserved for a long time even if their density is different from that of water, and the fundamental cause of this is the slowness of sedimentation, particularly if unavoidable turbulence in the water is taken into account.

Suspensions and colloidal solutions

A mixture which is homogeneous in the sense we have just described is called a *suspension* or a *colloidal solution*. The word 'solution' on its own is reserved for cases where the solute is a small molecule or ion, although the distinction is clearly not a sharp one since molecular sizes cover an enormous range up to the macromolecules of the previous chapter. However, a distinction can be made by using the technique of *dialysis*. Certain membranes are porous but have a very small pore diameter of around 2 nm: a sheet of parchment is one example but there are others from a variety of materials. If such a membrane is placed between a solution (inner compartment in Fig. 10.7) and pure water (outer compartment), it will allow molecules less than 2 nm across to pass through it but will stop particles or molecules larger than that. If the liquid within the inner compartment is a true solution, then complete homogenization will take place until there is the same concentration on both sides of the membrane. If, at the other extreme, the inner compartment contains a colloidal solution, it will remain intact. Finally, however, *a solution containing a mixture of large molecules and mineral salts in the inner compartment will lose the salts to the water outside.* This last case is used in the dialysis of blood to eliminate waste products such as urea (a small molecule of eight atoms) without altering the content of vital elements in the form of macromolecules and corpuscles. In other words, the process does what the kidney does when working normally.

Some examples of colloidal solutions

These fall into two main groups. In the first, the particle is a *chemically well-defined macromolecule* so that all the particles in the solution are identical. The white of an egg, for example, is to a first approximation a 12% colloidal solution of ovalbumin, a protein whose molecule has the shape of an ellipsoid of revolution with axes of 3.5 and 11 nm. Figure 10.8 gives an indication of the shapes and sizes of several proteins which can form colloidal solutions, together with some molecules and ions that form true solutions on the same scale for comparison.

In the second group of colloidal solutions, the particle consists of *a collection of atoms or small molecules* and is generally called a

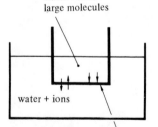

water + ions + large molecules

water + ions

membrane

Fig. 10.7 Principle of an experiment on dialysis.

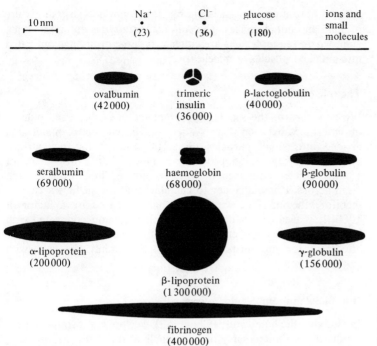

Na⁺
•
(23)

Cl⁻
•
(36)

glucose
▬
(180)

ions and
small
molecules

10 nm
⊢——⊣

ovalbumin
(42 000)

trimeric
insulin
(36 000)

β-lactoglobulin
(40 000)

seralbumin
(69 000)

haemoglobin
(68 000)

β-globulin
(90 000)

α-lipoprotein
(200 000)

γ-globulin
(156 000)

β-lipoprotein
(1 300 000)

fibrinogen
(400 000)

Fig. 10.8 General shapes and molar masses of several proteins compared with those of ions and small molecules.

micelle. Such a particle is produced either by the *coalescence* of isolated molecules or by the *dispersion* of a mass of material into smaller pieces: neither of these two opposing methods of formation leads to a fixed size for the particle. Examples of the first type of particle are provided by the micelles formed from soap molecules, which were described in the previous chapter. A typical example of the second type is provided by a dispersion of very small silver crystals produced by a spark discharge between silver electrodes in water: the crystalline particles produced are of varying sizes around a few nanometres and the liquid obtained forms the basis of a pharmaceutical product called collargel.

If the dispersed particles in a liquid come from *another liquid*, the mixture is called an *emulsion.* An emulsion of oil in water or water in oil is obtained by violently shaking up two separate layers of oil and water in a flask. Another example is milk, an emulsion of fat in water, whose white colour is the result of the scattering of light by micelles. Such scattering depends more on the size of the micelles than on their constitution.

Microemulsions can be produced in which the droplets of less than one micrometre are so fine that the medium remains transparent and appears to be homogeneous. Such tiny droplets are stabilized by a monomolecular layer of a substance called a *surface-active agent*, or *surfactant.*

There is a continuous gradation from the *ionic solution* to the *suspension of solid particles* about 0.01 mm across (an example of the latter being a suspension of clay in muddy water). The finest dispersions are perfectly stable while the coarsest are more or less slowly deposited by sedimentation: we pass from something we call without hesitation a single liquid phase to what is clearly a mixture of solid and liquid phases. The *colloidal solution* is an intermediate state where the molecules are either very large or are agglomerated into small corpuscles or micelles.

The role of interfaces

We now turn to the second important factor in dispersed phases, the interfaces between them. When a given volume of material is divided into small particles, the total area of its free surface becomes very large. Suppose, for instance, that a drop of oil of 1 mm diameter immersed in water is divided into droplets each 1 μm across. It is quite easy to calculate that the surface area, and therefore the interface with water, has increased by a factor of 1000. It is clear, therefore, that any phenomenon connected with interfaces, while completely negligible for large particles, becomes important for fine suspensions. This is certainly the case with *surface energy*.

Interfacial monomolecular layers

In studying the liquid state (Chapter 7), we spoke of *surface tension* produced by the tendency of a molecule at the surface to move to the interior. This is caused by the attractive force from neighbouring molecules which, for a surface molecule, are clearly all on one side of it (Fig. 7.8). The same effect occurs at the interface between two liquids.

Surface tension tends to reduce the surface area of a liquid as if it were enclosed in a stretched rubber sheet. If the shape of a given volume of liquid is made to change in such a way that the surface area is increased, an input of energy is required since some molecules must be transported from the interior to the surface and this requires work to be done. When the volume of liquid is very large, the surface energy does not affect the equilibrium conditions in practice because it is negligible compared with variations of energy due to gravity (p. 145). It is quite different, however, for fine dispersions.

Consider an emulsion of oil in water. We have seen why this might be stable if the droplets are small enough, but to create the droplets in the first place out of the originally separated layers of oil and water requires energy to increase the interfacial area. Conversely, if the emulsion already exists and two droplets of oil come into contact, their coalescence would reduce the surface area and be advantageous from the energy point of view. Coalescence of oil drops therefore takes place quite readily: the emulsion is quickly destroyed and the original separated layers of oil and water are restored. Hence, to facilitate the creation of the emulsion and make it stable, *there must be a reduction in interfacial surface tension*. This can be done with products called **surface-active** agents.

We have already met some of these: the soaps of the previous chapter. There are others known by the name of **detergents.** Detergent molecules are similar to those of soap, i.e. they are generally elongated with a polar hydrophilic head and a non-polar hydrophobic tail. When introduced into water, any such molecules arriving at the air–water interface in the course of their motion remain there. Eventually, they cover the surface with a *monomolecular layer*, in which they are packed together with each molecule normal to the surface, its hydrophilic head towards the water and its tail pushed into the air (Fig. 10.9).

How was it ascertained that the layer had a structure like that? Not with microscopes of extraordinary power, but by very simple experiments such as the one we now describe, where a mercury surface is used instead of water since it is more spectacular. A bath of very clean mercury is prepared and its surface sprinkled uniformly with a fine powder of talc. A very dilute solution of some surface-active agent in a volatile solvent is made up and a drop of it is placed on the mercury surface. The talc around the drop is immediately repelled, leaving a clearly-defined area on the surface where the mercury is very shiny. It is over this area that the surface-active molecules have spread. Since the cross-section of one molecule is known and the total number of them can be worked out from the amount of solution used, the area that would be covered by a monomolecular layer can be calculated and is found to agree with the area marked out by the talc. Confirmation is provided by measuring the thickness of the layer using optical methods which yield a value equal to the length of the molecule (known from its formula).

Why does the monomolecular layer reduce the surface tension? We have already seen from Fig. 7.8 that the tension arises from the attraction of the molecules in the surface towards those in the interior. However, when water, say, is covered by a monomolecular layer, the water molecules at the surface are now also attracted by the molecules of the surface-active material, so that the action of the internal molecules is balanced. In addition, the surface of a drop has less tendency to diminish in area since that would mean that the molecules of the monomolecular layer would have to move into the interior of the water in a 'less comfortable' position: the hydrophobic end would no longer be protected from contact with water.

In the case of oil and water, too, detergents form monomolecular layers at the interface, the oil being compatible with the hydrophobic parts of the molecules. *Surface-active substances thus*

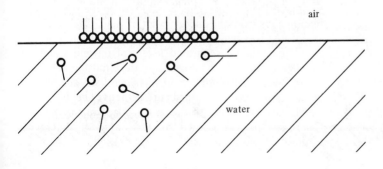

air

water

Fig. 10.9 Formation of a monomolecular layer of detergent on the surface of water. The molecules have hydrophilic heads (the open circles) and hydrophobic tails.

completely change the nature of interfaces between pure liquids and they do it by the addition of *extremely small amounts in proportion to the total volume*. Thus, the production of a monomolecular layer of soda soap on the surface of relatively large water droplets (1 mm diameter) needs a concentration of only one part in 10^8. Even if the drops were much smaller, and even if most of the soap remains in the interior of the water forming micelles (p. 202), a soap concentration of less than one part in 1000 is still enough to stabilize an oil–water emulsion.

The properties of the final product can be varied considerably by using quite subtle changes to modify the interfaces, changes such as altering the nature of the surface-active molecules introduced, or varying the temperature or the method of preparation. Thus, while a soda soap added to a mixture of oil and water gives an emulsion of oil in water, a calcium soap added to the same mixture gives an emulsion of water in oil having a much greater viscosity. Again, mayonnaise is an emulsion of oil in egg yolk and it does not take very much to produce a 'failure' of two superposed layers, one of which is the oil, rather than a perfectly smooth homogeneous sauce.

Wetting power

Pure water does not wet some surfaces: a duck's feathers, for instance, are always dry when it emerges from a pond. Similarly, if a little water is poured on to a lightly greased surface, it forms drops that run around quite easily. The drop owes its cohesion to surface tension since the water molecules prefer to remain with each other rather than adhere to the solid.

However, if the water contains a surface-active agent, the surface of the drop is covered by a layer of molecules with hydrophobic heads pointing outwards and able to adhere to the solid. Since the surface tension is then very small, the reduction in gravitational energy produced by flattening out becomes the dominant factor and the drop therefore spreads over the surface (Fig. 10.10). A number of drops behaving in this way can join up so that the solid surface becomes completely wet. For this reason, surface-active materials are also called **wetting agents**.

There are numerous applications of such agents. As one example, the liquid sprays that are used to treat agricultural crops would normally result in small droplets collecting on the leaves, but the addition of a wetting agent causes them to spread out over a large area with a resultant increase in efficiency.

The wetting of surfaces is a typical example of the way in which the very obvious behaviour of matter is a direct consequence of the structure at the molecular level. It also shows how a knowledge of this structure can yield very simple explanations of everyday occurrences.

Fig. 10.10 A spherical drop of pure water keeps its shape on a lightly greased surface; if the water contains a surface-active material, the drop spreads out considerably and the surface becomes wet.

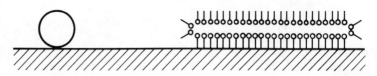

Detergent or cleaning action

Soiled fabrics are dirty because of solid particles of the 'dirt', often greasy, that cling to the fibres. They are washed using soaps or synthetic detergents, of course—but what is the mechanism that transforms dirty washing to clean? First of all, as we have just seen, the water containing the soap or detergent is capable of wetting the cloth: in other words, a layer of water covers each of its fibres. More exactly, as we saw in Fig. 10.10, there is a layer of detergent molecules adhering to the washing by their hydrophobic tails while the polar end is in contact with water molecules. This layer also covers the particles of dirt and can even insert itself between the particles and the fibre in the form of a layer folded on itself (Fig. 10.11). The cleaning action consists of: (a) the detachment of the particle from the fibre, (b) the particle itself covered in a monomolecular layer of detergent molecules, and (c) the subsequent removal of the particle if the water is agitated or if the fabric is rubbed. The particle covered by detergent molecules does not get deposited on the washing again, particularly if it is carried away by contact with cleaner and cleaner rinsing water. (In the process of *dry cleaning*, a liquid is used that *dissolves* the grease to form a true solution, and it is particularly important in that case not to let the solution evaporate in contact with the clothes since the dirt would then be redeposited on them.)

Foams

Soda water or beer in closed bottles contains dissolved carbon dioxide gas (p. 151). On opening the bottle, the liquid returns to atmospheric pressure and the gas is freed to form bubbles which, being very light, rise to the surface and often burst. However, with soapy water or with beer to which a surface-active agent has been added, the inner surfaces of the bubbles are covered by a monomolecular layer as is the free surface of the liquid. As the bubble approaches the surface, a film of the liquid is formed between the two monomolecular layers (Fig. 10.12). The film can become very thin indeed without breaking and in soapy water forms the 'soap bubbles'. The forces which determine the form of such films are the weight of the liquid and the surface tension, but the latter usually predominates because the weights are so small. A single bubble is therefore spherical. Several bubbles in contact are bounded by curved surfaces whose shapes are determined entirely by the equilibrium between the tensions in the surfaces: this is what we call a **foam**.

The sizes of the individual bubbles in a foam can vary a great deal. If they are small, the walls of film between them are very numerous and reflect incident light very strongly. This makes the foam appear white. Foams are never completely stable since the liquid eventually separates from the gas, but they can persist for some time. The head on a pint of beer, for instance, can last longer than ten minutes and we have all seen, unfortunately, the long-lasting foams created in rivers by detergents brought down with the wastewater from houses and factories.

Both the ability to produce foams and the cleaning power of a

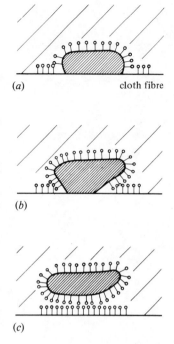

(a)　　　　　　　　　　　cloth fibre

(b)

(c)

Fig. 10.11 Showing how a detergent works—the particle of dirt stuck to the cloth is progressively surrounded by a monomolecular layer of detergent molecules, detached from the cloth and probably removed by the rinsing water.

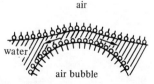

air

water　　　air bubble

Fig. 10.12 Formation of a thin layer surrounding an air bubble in water containing a surface-active substance.

given substance are consequences of the same underlying cause: the formation of monomolecular layers on surfaces. That does not mean, however, that one is a measure of the other: the best detergent is not, as is often believed, the one that produces most foam.

Interactions between micelles in a colloidal solution

Until now, we have neglected interactions *between particles* in our examination of the stability of colloidal solutions. This is not valid unless their concentration is very weak and, in fact, it is these very interactions that explain many of the phenomena observed in such solutions.

The most important fact is that *the micelles of a colloidal solution are electrically charged*, just as the ions in a true solution are carriers of positive or negative charges. This arises in one of two ways: either by dissociation of ions from the molecules forming the micelle, the ions being dispersed in the water, or by ions that are already in the water becoming attached to the micelle. Given the number of molecules that form a micelle, the number of elementary charges on each micelle can be quite considerable and can vary from one to another. In addition, the charge can be positive or negative according to the type of micelle and the composition of the liquid in which they are dispersed: there is no simple rule. The existence of the charges on micelles can be demonstrated by *electrophoresis*: a potential difference established between two electrodes immersed in the solution causes the micelles to move in a direction depending on the sign of their charge, just as the charged ions in an electrolyte move under the influence of an electric field.

A charged micelle creates an electric field in its neighbourhood so that ions of opposite sign in the liquid are attracted to it. An 'atmosphere' of these ions is formed which, at least partially, cancels the effect of the charge on the micelle and reduces its long-range influence—to zero if the assembly of micelle plus atmosphere is electrically neutral. Some of the ions in the atmosphere are stuck to the micelle, forming a double layer of charges around it, while others are free in the solution and mixed up with ions of opposite sign. It can be seen that *the electrical condition of a colloidal solution is complex, but it is one of the essential features that determine its behaviour.*

From what we have said, it is clear that two micelles have little or no interaction when they are separated by a distance at least equal, say, to their diameter. However, if in the course of their random thermal motion they should happen to approach more closely than this, their ionic atmospheres would begin to penetrate each other. In that case, the action between the micelles themselves would not be so exactly compensated and they would repel each other since they have charges of the same sign. This would therefore prevent the particles remaining in contact and would contribute to the stability of the colloidal solution.

Until ... yet another effect becomes involved. This is the Van der Waals attraction which exists between micelles just as much as between any other molecules (p. 75). It is quite distinct from the electrostatic interaction between charges and its main characteristic is that it is not significant except at very small distances. If the micelles are kept apart by their electrostatic repulsion, Van der Waals forces do not become involved and they remain isolated. If, however, they should come nearly into contact by random collisions *in spite of* the electrostatic repulsion, the Van der Waals attraction would outweigh the repulsion and the particles would remain stuck together. Others would come to join them. *This is what happens in the destruction of a colloidal solution by* **flocculation**: a mass of agglomerated micelles in dense clusters is formed, separated by clear liquid.

Flocculation can be easily produced by adding an electrolyte to a colloid: a solution containing a few hundredths of a mole per litre of NaCl would do, for instance, or one containing less than one-thousandth of a mole per litre of $CaCl_2$. The ions that are added, especially if they are divalent, reduce the thickness of the 'atmosphere' around the micelles, which can then approach each other sufficiently closely for the Van der Waals attraction to keep them together. The ion concentration of a colloidal solution can thus be used as a control: it allows us to make the solution more stable or to cause flocculation, whichever we need. This is frequently used in a number of processes in the chemical industry. It also explains the formation of a delta at the mouth of a river: the silt-laden water of the river is a colloidal solution which flocculates when it comes into contact with the salt water of the sea, and the large particles of mud which are formed are then rapidly deposited.

In some cases, flocculation is reversible: the cluster of micelles can be dispersed and put back into solution by the action of chemical substances called *peptizing agents*. In other cases, however, the system sets quite irreversibly into a firm mass: this is called **coagulation**. The micelles which are linked to each other then form an irregular network into which the water molecules are inserted and, in a sense, immobilized. The whole mass has the appearance of a very soft solid and is called a **gel**.

A gel is typically *elastic*: under the action of an external force, the micelles do not slide over each other because of the bonds between them which form anchorage points very reminiscent of those described as points of attachment in rubber (p. 189). Clearly, however, these attachments will give way under much smaller forces than will those in rubber.

White of egg, a colloidal solution of albumin, coagulates by heating at 80 °C to form a gel insoluble in water (hard egg white). *Gelatin*, a protein extracted from bones and skins of animals, *pectin*, which occurs in fruit, and *agar-agar*, extracted from sea-weed—all of these give a gel with water that is commonly called a *jelly* in Britain or a *jello* in the United States of America. What is surprising is that a gel which has the appearance of a solid, even if rather a soft one, contains a considerable proportion of water. A mere 1% concentration of macromolecules is sufficient to form a skeleton, solid enough to give the whole mass a firm consistency.

A gel can easily absorb water or lose it. A piece of solid gelatin in water absorbs it and swells up. Conversely, a gel of hydrated silica can be dehydrated to become a transparent solid with a glassy appearance called 'silica gel'. This has widespread use as a drying agent because it can absorb large quantities of water without changing significantly in appearance. The water can be driven off again by heating.

The bonds between micelles are not very strong, so that it is sometimes possible to break them simply by powerful disturbance of the medium. The gel then becomes a fluid colloidal solution once again, with the micelles dispersed. If left for a few moments, however, the bonds between micelles re-form and the material once more takes on the consistency of a gel. This reversible property is called *thixotropy* and is possessed by some paints which can be made to flow more readily for a short time if they are sufficiently disturbed. Thus they form a gel in the pot, but spread easily when applied with a brush or roller.

In conclusion:
a quick look over the whole subject

We have completed a long voyage of exploration through the structures of matter on an atomic scale. On the way, we have encountered some structures of great complexity and variety that do not seem to belong to the straightforward classification implied by the simple models used at the beginning of our study. For that reason, it now seems the right time to reconsider the whole subject from a point of view that is wide enough to ignore the details and lay bare the line of thought unifying the really essential features.

The basic idea in the early chapters was that a given substance, whether composed of atoms, ions or molecules, exists in two forms: one ordered, the crystalline state, and one disordered, the liquid state (confining ourselves to equilibrium forms of condensed matter). Only a discontinuous transition is possible between these two states: for instance, abrupt melting at a fixed temperature. Such a scheme is in fact observed in a large number of substances and the only thing that varies from one to another is the temperature of the transition.

In Chapter 9, however, we gave examples that do not conform to such a simple system of two states with well-defined properties, so that, if the initial idea is not applicable in every case, we are faced with the following question: what are the characteristics of substances that cause them to fall into the simple two-state framework or to fall outside it? The answer is in two parts, the second being rather more fundamental.

First of all, in order to belong to the simple two-state category, it is essential that what we have called the 'basic particle' of the structure retains its full identity as the temperature varies. It must not, for instance, be transformed in any way at higher temperatures, which must only affect the mutual arrangement of the particles that are themselves unchanged. This is clearly the case with isolated atoms and ions and with small molecules such as H_2O. However, some molecules and complex ions are quite fragile: an increase in the amplitude of thermal vibrations can cause internal bonds to be broken or changed at temperatures below that at which the crystalline structure would melt. In other words, the solid decomposes before melting and *the liquid phase does not exist*. As examples, we may quote calcite ($CaCO_3$), which decomposes into lime and CO_2, gypsum ($CaSO_4 . 2H_2O$), which dehydrates on heating to give plaster of Paris, and diamond, which transforms into graphite at high temperatures. Finally, numerous organic molecular crystals decompose before melting because of the extreme weakness of the bonds. In short, there are *chemical* reasons why some substances are known to exist only in the solid state.

A second type of deviation from the simple two-state scheme arises when the ordered state no longer exists or, rather, exists in a very imperfect form: it is replaced by intermediate states, examples of which were given in Chapter 9. It is easy to see why states between order and complete disorder exist. If a collection of molecules is to become ordered into a perfect crystal (or at least a 'good' crystal), all the molecules must be identical. This is normally the case with small molecules and even with some large ones where the composition and shape are absolutely fixed (insulin, Fig. 1.3, for example). However, this is not the case for molecules having a carbon chain in which successive C—C bonds are articulated, as in lipids (p. 201), for example, or in polymer macromolecules where, in addition, the length of the chain is not exactly fixed.

Such molecules cannot form a regular compact array nor, as a consequence, a good crystal. The perfectly ordered state, therefore, cannot exist. Neither, however, can a state of disorder exist which is as great as that in a liquid of small spherical atoms. The presence of chains and the chemical bonds between them produce a certain degree of ordering.

From this, we see that states of intermediate order are observed in substances with *flexible molecules of great size*. The simple two-state scheme of good crystal and normal disordered liquid is therefore restricted to substances with small and rigid molecules.

In the case of liquid crystals (p. 192), there is also the particular elongated cylindrical shape of the molecules which brings about the existence of mesophases. At low temperatures, such molecules are capable of becoming ordered into a crystalline array because of their moderate length and their interactions, while at high temperatures (~ 200 °C), they are disordered, with the structure of a normal liquid. It is at intermediate temperatures that phases of partial order exist, such as those we described in Chapter 9.

We can thus summarize this quick review by saying that the scheme of two well-defined states (order and disorder) covers the structures of minerals in general and of other substances with small molecules, with the proviso that the disordered phase may not exist because of chemical instability. *The states of intermediate order, however, are encountered among substances with large flexible molecules, especially those with long carbon chains.*

Here we have put our finger on the reason why the study of intermediate states is important: they form the basis for our understanding of the physical properties of many essential materials: on the one hand of polymers that form plastics, and on the other of matter that forms living beings. Biochemical studies show that many constituents of living organs are composed of large molecules, very often with long chains. From what we have been saying, we can predict that these substances have intermediate structures. It is reasonable to suppose that the physiological properties of the organs depend on their structure on an atomic scale as well as on chemical reactions between their constituents. Colloids, a state of matter intermediate between solutions of small molecules and suspensions of microscopic particles, have long been known to play an important role in living organisms.

It is clear that to determine the quite irregular structure of a biological substance presents a problem of great complexity that has, until now, only been touched on. The description of a partially ordered structure is very difficult because of the large number of parameters involved. In addition, experimental techniques, such as X-ray diffraction and electron microscopy, are much less useful for disordered structures than they are for crystals. There are, however, several reasons for being confident. We must not forget that the determination of the structure of a crystalline protein, which is common enough today, was considered as insoluble thirty years ago; moreover, images of atoms themselves have very recently been obtained (Fig. 4.22); finally, there are major advances in techniques in the offing, such as X-ray sources 1000 times more powerful than previous ones that are now beginning to be available.

Following the great success of research into the structures of the ordered state and following similar success with disordered structures, the field of study covering states of intermediate order is just opening up. The work will be long and hard, but the stakes are high because the structures are linked to the way in which living organisms function.

Index